Fungi in vegetation science

Handbook of vegetation science

FOUNDED BY R. TÜXEN
H. LIETH, EDITOR IN CHIEF

Volume 19/1

Fungi in vegetation science

Edited by

W. WINTERHOFF

Pädagogische Hochschule Heidelberg, Heidelberg, Germany

Springer-Science+Business Media, B.V.

Library of Congress Cataloging-in-Publication Data

Fungi in vegetation science / edited by W. Winterhoff.
p. cm. -- (Handbook of vegetation science ; v. 19/1)
Includes bibliographical references and index.
ISBN 978-94-010-5065-4 ISBN 978-94-011-2414-0 (eBook)
DOI 10.1007/978-94-011-2414-0
1. Fungal communities. I. Winterhoff, Wulfard. II. Series:
Handbook of vegetation science ; pt. 19/1.
QK911.H3 pt. 19/1
[QK604.2.C64]
581 s--dc20
[589.2'045247] 92-7037

ISBN 978-94-010-5065-4

Contents

Series Editor's preface

Of all taxa of any ecosystem, the lower plants, the soil microflora and fauna are the least understood.

Vegetation scientists have been pre-occupied so much with the distribution pattern of higher plants that they have paid very little attention, if any, to the lower plants. It is the specific objective of volume 19 of the Handbook of Vegetation Science to compile all current knowledge and in this way demonstrate the gaps in our knowledge and also show the potential we can expect if we devote more attention to the distribution pattern of lower plants.

The number of scientists interested and competent in phytosociological and ecological matters of lower plants is apparently not in proportion to the importance of such problems. The fact that we did not succeed in getting parts together for a volume on mosses and lichens proves this fact quite clearly.

From this position we decided to publish volume 19 in individual sections accordingly as they become available. We expect now to deal with this problem in some greater detail as we uncovered an obvious need for better information in this area and consequently the need for separate editors for each section.

We start volume 19 with a first section on fungi. Prof. Winterhoff succeeded in gathering a group of competent specialists on distribution patterns of fungi in ecosystems. The volume concentrates on macrofungi and some aspects of microfungal associations. The problems and requirements for further actions are well elaborated by Prof. Winterhoff and the contributors to this volume.

We offer this section to the scientific community in the hope that it will be useful for field work and at the same time will trigger more interest in the forthcoming sections for mosses, lichens and lower plant coenological considerations in other regions of the world.

Osnabrück, 18th May 1992 H. LIETH

List of contributors

E. J. M. Arnolds, Biologisch Station te Wijster, afdeling Vegetationskunde, Kampsweg 27, NL-9418 PD Wijster, The Netherlands.

U. Braun, Martin-Luther-Universität, FB. Biologie, Institut für Geobotanik und Botanischer Garten, Neuwerk 21, O-4010 Halle/S, Germany.

A. Bujakiewicz, Adam Mickiewicz University, Department of Plant Ecology and Protection of Environment, Al. Niepodległości 14, PL-61-713 Poznán, Poland.

W. Gams, Centraalbureau voor Schimmelcultures, P.O. Box 273, NL-3740 AG Baarn, The Netherlands.

G. Hirsch, Stauffenbergstraße 4, O-6902 Jena-Lobeda, Germany.

G. Kost, Universität Tübingen, Institut für Biologie I, LB Spezielle Botanik, Auf der Morgenstelle 1, W-7400 Tübingen 1, Germany.

M. Lisiewska, Adam Mickiewicz University, Department of Plant Ecology and Environment Protection, Al. Niepodległości 14, PL-61-713 Poznán, Poland.

W. Winterhoff, Pädagogische Hochschule Heidelberg, Keplerstraße 87, W-6900 Heidelberg 1, Germany.

1. Introduction

W. WINTERHOFF

Readers will perhaps be surprised to find a volume about fungi within a handbook of vegetation science. Although fungi traditionally feature in textbooks on botany, at least since Whittaker (1969), they have mostly been categorised as an independent kingdom of organisms or in contrast to the animal and plant kingdom as probionta together with algae and protozoa. More relevant for ecology than the systematic separation of fungi from plants is the different life style of fungi which in contrast to most plants live as parasites, saprophytes or in symbiosis. Theoretical factors aside, there are also practical methodological considerations which favour the distinction between fungal and plant communities, as has been shown for example by Dörfelt (1974).

Despite the special position of the coenology of fungi the topic is included here in the handbook of vegetation science. We realize that important differences exist between fungal and plant communities. The reasons for including the former are that mycocoenology developed from phytocoenology, the similarity of the methods and concepts still employed today and the close correlation between fungi and plants in biocoenoses require a treatment of mycocoenology within the scope of a handbook of vegetation science.

In contrast to some authors (e.g. Meisel-Jahn and Pirk 1955) who view fungi as elements of plant communities, we, along with Hueck (1953), Darimont (1973), Kreisel (1957) and many more, approach the communities of fungi within biocoenoses separately from the phytocoenoses as mycocoenoses or mycosocieties (see chapter 2).

Mycocoenology is concerned with fungal communities. We see these as consisting, on the one hand, of mycocoenoses (as referred to by Barkman 1976 and Arnolds 1981), i.e. the fungal elements in the individual biocoenoses and, on the other hand, of mycosocieties, which form within mycocoenoses particularly in special microhabitats (see chapter 2).

W. Winterhoff (ed.), Fungi in Vegetation Science, 1–5.

According to Arnolds (1981) mycocoenology can be defined as "those investigations dealing with comparative, more or less quantitative, analysis of mycocoenoses or mycosocieties in selected plots or transects in order to establish relations between myco- and phytocoenoses or to describe mycocoena." In terms of its objectives and the questions it poses, mycocoenology is a subdiscipline of both mycoecology and biocoenology.

The so-called mycofloristic-ecological approach (Arnolds 1981) will also be taken into account in this volume as a further subdiscipline of mycoecology. For information on the autecology of fungi, however, textbooks and handbooks on mycoecology should be consulted, e.g. Agerer (1985), Cooke (1979), Cooke and Rayner (1984), Frankland *et al.* (1982), Harley (1969), Harley and Smith (1983), Jahn (1990), Rayner and Boddy (1988) and Rypacek (1966).

Mycocoenology is still a relatively young science, its development starting with the work of Haas (1932) and Höfler (1937), which has been described, amongst others, by Arnolds (1981). Although nowadays work on mycocoenology is carried out in most European countries, especially in Poland and the Netherlands, even here most field mycologists are still exclusively concerned with taxonomical and chorological questions. Outside Europe relatively little work has been undertaken in this field. This is due, on the one hand, to methodological problems, on the other hand to insufficient knowledge about fungal flora even in Europe (cf. Barkman 1987).

At present, the great number of competing approaches, sampling methods and formation of concepts in mycocoenology, although enabling individual problems to be worked out thoroughly, do make it difficult to compare and synthesize the results. Even the central term 'mycocoenosis' is used in different meanings (see chapters 2, 6 and 8).

The coenology of microfungi has, to a great extent, developed independently from that of macrofungi, because of the difference in the type of methods required (see chapters 1.2 and 6.1). Despite the attempts made by Apinis (1972) to discourage mycologists from restricting their studies of fungal communities to specific groups, little research has, to this day, been undertaken which has dealt with microfungi and macrofungi in the same way. This volume will also therefore present the coenology of microfungi and that of macrofungi in separate chapters.

Apart from the systematic position and form of nutrition there are considerable differences between phytocoenoses and mycocoenoses: usually the mycelia live totally inside the soil or other substrates; structure and species combination are not determined by competition for light; in forests mycocoenoses usually differ from phytocoenoses through their greater number of species, their larger minimal area (Winterhoff 1975, Arnolds 1981) and through greater fluctuations. These differences can at least partly be attributed to the heterotrophic nature of fungi. Due to their ability to specialise in certain substrates, hosts or partner plants, in forests fungi have a far greater niche differentiation than plants, thus enabling many more species to live in the same mycocoenosis (cf. Barkman 1987). Considering the small surface area which is usually taken up by hosts, partner plants, deadwood and other special substrates, the percentage of the whole area which could be potentially occupied by specialising fungal species is much smaller again. Presumably, the likelihood of such fungi establishing themselves is therefore much lower in turn. The speed at which infected plant matter dies and decomposes forces many parasites and saprophytes to seek out new habitats more frequently than most forest plants are forced to do.

The heterotrophic nature of fungi means that they are either directly or indirectly dependent on green plants. However, they are not necessarily dependent on certain plant communities. In some circumstances, the existence of one suitable host plant is sufficient irrespective of the plant community in which it is found or whether it is isolated for example as a tree by the roadside. In special cases, fungi can also live as 'aliens' in biocoenoses where there are no plants producing the required substrate or even in plant-free environments such as caves or buildings, as long as the necessary substrate has been transported there.

Despite these exceptions most mycocoenoses and mycosocieties tend to be closely connected with certain plant communities. This is because most host and partner plants associated with fungi are not scattered around arbitrarily but only live in certain plant communities. The occurrence of fungal and plant species depends largely on the same climatic and soil conditions and important local factors, such as humidity, are partially determined by plant communities.

On the other hand, although fungi and their communities are dependent on plants they are still necessary for the existence of most phytocoenoses and also affect their composition. Fungi contribute greatly to the formation of humus and to the remineralisation of remaining debris, particularly the decomposition of deadwood. The most important forest trees need ectomycorrhizal fungi in order to flourish, and plant vitality as well as their ability to compete with each other can be greatly influenced by parasitic fungi (see chapter 8).

Mycocoenology has far-reaching implications for other areas outside the field of mycoecology. As essential compartments of most ecosystems are largely formed by fungi, it is absolutely necessary for our understanding of the ecosystems that we record their mycocoenoses and research the links these have to other compartments in the ecosystems. Communities of wood-rotting fungi in particular are of practical importance for forestry and the timber industries. As has been shown, for example, by Winterhoff (1975), Arnolds (1981) and Bujakiewicz (1987), fungi as differential taxa lend themselves to a detailed subdivision of plant communities in many cases and to provide a synopsis of corresponding units of vegetation in different regions of flora. Mycocoenoses reflect very closely the state of the ecosystems. For example, they reveal more accurately than plants some local factors such as the type of organic matter and the type of fertilization (Arnolds 1981). Any change in the mycocoenoses in grasslands and forests could be the first indication of an imminent change in the system as a whole, as in forest decline for example.

Today mycocoenology gains particular significance for the conservation of endangered fungi. In many European countries there has been a substantial and rapid drop in the amount of fungal flora over the last thirty years (cf. e.g. Arnolds 1985, 1991, Benkert 1982, Derbsch and Schmitt 1984, Fellner 1983, Jansen 1989, Nitare 1988, Winterhoff 1984 and Wojewoda and Lawrynowicz 1986). As changes in and the destruction of the biotopes and vegetation are the main causes of the decline, those particular plant communities and substrates on which the endangered fungi rely need to be conserved in order to protect the fungal flora. One of the prerequisites, therefore, will be to establish and record the links between fungi and plant communities as well as substrates.

I would like to thank Christina Lausevic for the above translation.

References

Agerer, R. 1985. Zur Ökologie der Mykorrhizapilze. Bibliotheca mycologica 97.

Apinis, A. E. 1972. Facts and problems. Mycopath. Myc. Appl. 48: 93—109.

Arnolds, E. 1981. Ecology and coenology of macrofungi in grasslands and moist heathlands in Drenthe, the Netherlands. Bibliotheca Mycologica 83.

Arnolds, E. (red.) 1985. Veranderingen in de paddestoelenflora (mycoflora). Wetenschappelijke mededelingen K.N.N.V. 167: 102 pp.

Arnolds, E. 1991. Decline of ecotomycorrhizal fungi in Europe. Agriculture, Ecosystems and Environment 35: 209—244.

Barkman, J. J. 1976. Algemene inleiding tot de oekologie en sociologie van macrofungi. Coolia 19: 57—66.

Barkman, J. J. 1987. Methods and results of mycocoenological research in the Netherlands. p. 7—38. *In*: G. Pacioni (ed.) Studies on fungal communities. University of Aquila, Italy.

Benkert, D. 1982. Vorläufige Liste der verschollenen und gefährdetern Großpilzarten der DDR. Boletus 6: 21—32.

Bujakiewicz, A. 1987. Indicative value of macromycetes in the forest associations of Mt. Babia

(S. Poland). p. 41—47. *In*: G. Pacioni (ed.) Studies on fugnal communities. University of Aquila, Italy.

Cooke, W. B. 1979. The ecology of fungi. CRC Press Inc., Boca Raton, Florida.

Cooke, R. C., and A. D. Rayner. 1984. Ecology of saprotrophic fungi. Longman, London.

Darimont, F. 1973. Recherches mycologiques dans les forets de Haute Belgique. Essai sur les fondements de la sociologie des champignons supérieurs. Mém. Inst. Roy. sc. nat. Belg. 170.

Derbsch, H., and J. A. Schmitt. 1984. Atlas der Pilze des Saarlandes. 1. Verbreitung und Gerfährdung. Aus Natur und Landschaft im Saarland. Sonderband 2, Saarbrücken.

Dörfelt, H. 1974. Zur Frage der Beziehungen zwischen Mykocoenosen und Phytocoenosen. Arch. Natursch. Landschaftsforsch. 14: 225—228.

Fellner, R. 1983. Mycorrhizae-forming fungi in climax forest communities at the timberline in Giant Mountains. Ceská mykol. 37: 109.

Frankland, J., J. N. Hedger, and M. J. Swift (eds.). 1982. Decomposer basidomycetes. Their biology and ecology. Brit. Mycol. Soc. symposium ser. 4.

Haas, H. 1932. Die bondenbewohnenden Großpilze in den Waldformationen einiger Gebiete von Württemberg. Beih. bot. Centralbl. B 50: 35—134.

Harley, J. L. 1969. The biology of mycorrhiza (2nd ed.). Leonard Hill Books, London.

Harley, I. L., and S. E. Smith (eds.). 1983. Mycorrhizal symbiosis. Academic Press, London.

Höfler, K. 1937. Pilzsoziologie. Ber. dtsch. bot. Ges. 55: 606— 622.

Hueck, H. J. 1953. Myco-sociological methods of investigation. Vegetatio 4: 84—101.

Jahn, H. 1990. Pilze an Bäumen (2nd ed.). Patzer, Berlin, Hannover.

Jansen, A. E. 1989. De eerste bijeenkomst van het Europees Comité voor de bescherming van paddenstoelen. Coolia 32: 11—13.

Kreisel, H. 1957. Die Pilzflora des Darß und ihre Stellung in der Gesamvegetation. Feddes Rep. Beih. 137: 110—183.

Meisel-Jahn, S., and W. Pirk. 1955. Über das soziologische Verhalten von Pilzen in Fichten-Forstgesellschaften. Mitt. Flor. - soz. Arbeitsgem. N.F. 5: 59—63.

Nitare, J. 1988. Jordtungor, en svampgrupp pa tillbakagang i naturliga fodermarker. Svensk bot. tidskr. 82: 341—368.

Rayner, A. D. M., and L. Boddy. 1988. Fungal decomposition of wood, its biology and ecology. John Wiley & Sons, Chichester.

Rypacek, V. 1966. Biologie holzzerstörender Pilze. VEB Gustav Fischer, Jena.

Whittaker, R. H. 1969. New concepts of kingdoms of organisms. Science 163: 150—161.

Winterhoff, W. 1975. Die Pilzvegetation der Dünenrasen bei Sandhausen (nördliche Oberrheinebene). Beitr. naturk. Forsch. Südwestdeutschl. 34: 445—462.

Winterhoff, W. *et al.* 1984. Voläufige Rote Liste der Großpilze (Makromyzeten). p. 162—184. *In*: J. Blab, E. Nowak, W. Trautmann, and H. Sukopp Rote (eds.) Liste der gefährdeten Tiere und Pflanzen in der Bundesrepublik Deutschland. Naturschutz aktuell 1.

Wojewoda, W., and M. Lawrynowicz (1986). Red list of threatened macrofungi in Poland. p. 45—82. *In*: K. Zarycki, and W. Wojewoda (eds.) List of threatened plants in Poland. Polish Scientific Publishers, Warszawa.

2. The analysis and classification of fungal communities with special reference to macrofungi

EEF ARNOLDS

Summary

This chapter treats the various approaches, methods and terms, used for the study of fungal communities in general, and communities of macrofungi in particular. The morphological, taxonomic, ecological and methodological differences between macro- and microfungi are discussed. Communities of macrofungi are studied in different ways, using mycocoenological, mycosynusial, mycofloristic-ecological, phytocoenological and geographical approaches, which are concisely treated. The status of fungal communities, mycocoenoses and mycosocieties is discussed, as well as the related terminology. Attention is paid to the different degrees of dependence of fungi on other organisms and to the consequences for analysis and classification of mycocoenoses. The criteria for the selection of sample plots, viz. uniformity, size and representativity, are reviewed. The more important methods for qualitative and quantitative analysis of mycocoenoses are presented, as well as methods for the synthesis of these data in relation to frequency and duration of mycocoenological analysis. Finally, the different approaches to classification of mycocoenoses are critically considered.

2.1. Introduction

Fungi are important components of almost all the ecosystems performing a variety of ecological functions. The three main functional groups are (a) saprotrophic fungi, decomposing dead organic matter, (b) biotrophic fungi, living in mutual symbiosis with green plants (mycorrhizae, lichens) and (c) necrotrophic fungi, living as parasites on other organisms, comprising plants, animals and other fungi (§ 2.2.2). An assemblage of fungi in a certain uniform

W. Winterhoff (ed.), Fungi in Vegetation Science, 7—47.

habitat can be regarded and studied as a fungus community. These communities range in size from a few mm^2 on seeds (seed-born fungi, mainly Deuteromycetes, e.g. Neergaard 1977) and living beetles (mainly Laboulbeniomycetes, e.g. Benjamin 1973); a few cm^2 on living and senescent leaves (phylloplane fungi), herbaceous plants, twigs and animal excrements; in the order of dm^2 on branches and moss carpets, m^2 on logs and trunks and hundreds of m^2 on the forest floor (ectomycorrhizal fungi on living roots; saprophytes on litter). Well-developed fungal communities are also found on various substrates in salt and fresh water (e.g. Sparrow 1960; Ingold 1975).

Fungi are essentially micro-organisms. The vegetative structures (mycelia) cannot, with few exceptions, be perceived with the naked eye or identified in the field. Identification usually depends on the presence of reproductive organs (sporocarps, conidiophores) which are extremely variable in size, appearance and periodicity. Part of the species of various taxonomic groups produce conspicuous, macroscopic sporocarps ('toadstools, mushrooms') and are collectively known as macrofungi. The remaining lack such large reproductive structures (§ 2.2). It is not surprising that different methods have been developed for the study of communities of macrofungi, which are mainly derived from methods in plant ecology, and microfungi, which are based on microbiological techniques.

In the present chapter the analysis and classification of macrofungal communities will be discussed, whereas the methodological aspects of microfungal communities will only be stipulated (§ 2.2; see chapter 7). Special problems in the study of macrofungal communities, compared with most studies of vegetation, are (a) the inaccessibility of vegetative structures (mycelia) for research, hence dependence on sporocarps, (b) short duration of sporocarps (§ 2.8), (c) strong periodicity and fluctuations of fruiting (§ 2.8), (d) great variation in ecological function (§ 2.2.2), (e) taxonomic problems of reliable identification (§ 2.6). For more extensive discussions on these problems, see e.g. Höfler (1937), Hueck (1953), Kalamees (1968), Haas (1972), Darimont (1973), Arnolds (1981), Winterhoff (1984) and Kreisel and Dörfelt (1985).

The analysis of communities of macrofungi was previously treated in an adequate way in this Handbook (vol. 4) by Winterhoff (1984). It was considered useful to include a chapter on this subject in this volume as well because (1) a contribution on methodology cannot be missed in view of the other chapters on fungal communities; (2) important theoretical and practical progress was made since the manuscript by Winterhoff was completed and (3) the present chapter presents a different, more theoretical approach to the problem.

For surveys of the historical development of myco-ecological research the reader is referred to Darimont (1973) and Fellner (1987).

2.2. Macrofungi and microfungi

2.2.1. *Morphological and taxonomic criteria*

Macrofungi are those fungi forming reproductive structures (sporocarps; also called sporophores, carpophores, fruitbodies, mushrooms and toadstools etc.), which are individually visible with the naked eye, that is larger than about 1 mm (e.g. Arnolds 1981:15, 1984). The remaining fungi are called *microfungi*. The limits between the two groups are very arbitrary and do not coincide with taxonomic or ecological units (Table 1). In extreme cases even two species of a single genus can belong to macro- and microfungi, respectively: *Tremella encephala* Pers. has striking cushion-shaped basidiocarps up to 40 mm broad, but *Tremella obscura* (Olive) M.P. Christ. has a completely reduced thallus and lives as a parasite within the small basidiocarps of *Dacrymyces stillatus* Nees: Fr. (Torkelsen 1968). It can only be observed accidentally when *Dacrymyces* is studied under the microscope. On the other hand, all Myxomycotina are mostly neglected in studies on macrofungal communities, although some species of *Fuligo* Hall. and *Lycogala* Adans. have reproductive organs up to 100 mm broad.

Some groups of unmistakable macrofungi are nevertheless strongly underrepresented in most mycocoenological studies, due to their hidden sporocarps, especially hypogeous fungi (e.a. Tuberales, Plectascales p.p., Gasteromycetes p.p.) and resupinate Aphyllophorales growing at the underside of wood (e.g. De Vries 1990).

The delimitation of macrofungi is different from author to author, dependent on the attempted accuracy of research and the taxonomic knowledge. In Table I a survey is given of taxonomic groups that are either constantly or accidentally involved in studies of macrofungal communities. The application of a narrow or broad concept of macrofungi may considerably influence some results of coenological studies, for instance data on species diversity.

2.2.2. *Ecological aspects*

Both macrofungi and microfungi comprise saprotrophic, biotrophic, and necrotrophic fungi, but their ecological main points are different.

Table I. Taxonomic groups, assigned to the macrofungi according to mycocoenological studies

Division **MYXOMYCOTA**	(−−, pp. min)
Division **EUMYCOTA**	
Subdivision **Basidiomycotina**	
Class Hymenomycetes	
Order Agaricales	++
Order Russulales	++
Order Boletales	++
Order Aphyllophorales	
'Cantharelloid fungi'	++
'Hydnoid fungi'	++
'Poroid fungi'	++
'Clavarioid fungi'	+
'Stereoid fungi'	+
'Merulioid fungi'	+
'Corticioid fungi'	(±)
'Cyphelloid fungi'	(−)
Class Heterobasidiomycetes	
Order Auriculariales	(+)
Order Tremellales	(+)
Order Dacrymycetales	±
Order Septobasidiales	−−
Order Exobasidiales	−−
Order Tulasnellales	−−
Class Gasteromycetes	
Orders with epigeous sporocarps:	
Order Lycoperdales	++
Order Sclerodermatales	++
Order Tulostomatales	++
Order Phallales	++
Orders with hypogeous sporocarps:	
Order Gastrosporiales	−
Order Gauteriales	−
Order Hysterangiales	−
Order Hymenogastrales	−
Order Leucogastrales	−
Order Melanogastrales	−
Class Teliomycetes	
Order Ustilaginales	−−
Order Uredinales	−−
Subdivision **Ascomycotina**	
Class Hemiascomycetes	o
Class Loculoascomycetes	o
Class Plectomycetes	o
Class Pyrenomycetes	
Order Coronophorales	(−−)

Table I. (continued)

Order Sphaeriales	(±, p.p. min.)
Class Discomycetes	
Order Pezizales	(+)
Order Helotiales	(±, p.p. min.)
Order Tuberales	–
Order Clavicipitales	(–)
Order Plectascales	(––)
Order Lecanorales	(––)
Subdivision **Zygomycotina**	(––, p.p. min.)
Subdivision **Deuteromycotina**	(––, p.p. min.)

Meaning symbols:
++ : always considered
+ : mostly considered
± : in some studies considered, in others not
– : rarely considered
–– : very rarely considered
o : not considered
Symbols between brackets indicate that usually only part of the existing species is considered.

Among the saprotrophic fungi most microfungi are involved in the decomposition of simple carbon compounds, for instance mono- and disaccharids leaking from living roots and leaves and available in fresh dung, dead animals, etcetera. Furthermore numerous species are able to breakdown cellulose and hemicellulose, but only few are involved in lignin decomposition. Most saprotrophic macrofungi contribute substantially to the chemical degradation of cellulose, hemicellulose and lignin, the latter substance being almost exclusively tackled by basidiomycetes (in majority macrofungi). For surveys of the biology and ecology of saprotrophic fungi, the reader is referred to Dickinson and Pugh (1974), Swift *et al.* (1979), Frankland *et al.* (1982) and Cooke and Rayner (1984).

The biotrophic fungi living in symbiosis with green and blue algae are forming characteristic morphological structures known as lichens. Most of them are inoperculate ascomycetes with ascocarps in the range of 0.1—10 mm, which might be assigned, like e.g. the Helotiales, to the macrofungi or not (Table I). Only very few basidiomycetes (some Agaricales and clavarioid fungi) are associated with algae in basidiolichens. Their basidiocarps belong unmistakably to the macrofungi. Since lichens are in morphological and biological respect units of their own, they are not treated in this volume. Fungal and algal components of lichens are never studied separately in phytocoenology with the exception of basidiolichens. These species are usually neglected in the study of lichen communities, but included in myco-ecological work.

Many biotrophic fungi live in mutualistic symbiosis with living plant roots, forming different kinds of mycorrhizae (e.g. Harley and Smith 1983). The most wide-spread type is vesicular-arbiscular mycorrhiza, formed between the majority of land plants and phycomycetes, assigned to the microfungi (survey by Sanders *et al.* 1975). The second important type is ectomycorrhiza between fungi and many woody and some herbaceous plants. The fungi involved in ectomycorrhiza are mainly basidiomycetes with macroscopic basidiocarps, less often Asco- and Phycomycetes, which are usually also assigned to the macrofungi (e.g. Marks and Kozlowski 1973). Basidiomycetes are often also participating in more specialized mycorrhizae, such as orchid and arbutoid mycorrhiza (Harley and Smith 1983).

Necrotrophic fungi derive their organic nutrients from dead cells of organisms which they have themselves killed (Cooke 1977). Consequently, not the entire organism is necessarily killed. Most necrotrophic fungi are microfungi, or at least treated in practice as such, although some well-known plant pathogens (e.g. Uridinales and Ustilaginales) can be easily recognized macroscopically. Among the generally recognized macrofungi relatively few species are necrotrophic, most of them on woody plants, e.g. Agaricales, such as *Armillaria mellea* (Vahl: Fr.) Kumm. ss. str. and Aphyllophorales, such as *Heterobasidion annosum* (Fr.) Bref. Some species of ascomycetes, parasiting insects and fungi, are also usually included in the macrofungi, e.g. the genus *Cordyceps.*

It should be stressed that this division in functional groups is only schematical and that all kinds of intermediates and transitions occur (see also Kreisel and Dörfelt 1985). For instance, a necrotrophic fungus is often able to continue as a saprotroph after the death of its host. Fungi which are normally ecotmycorrhizal may behave as saprophytes in some cases. The balance between fungus and host in mycorrhiza may also shift to a more or less necrotrophic way of life. A good example is *Thelephora terrestris* Ehrh.: Fr., which is often ectomycorrhizal, but it may kill young trees in nurseries and a resupinate form lives as a saprophyte on dead wood (Corner 1968).

Conclusively, macrofungi include a considerable part of the saprotrophic fungi (mainly degrading cellulose, hemicellulose and lignin), most ectomycorrhizal fungi and some necrotrophic fungi.

2.2.3. *Methodological aspects*

The study of communities of macrofungi has been initiated by vegetation scientists and agaricologists, which is reflected in the applied methods: they show a lot of resemblance to methods for the study of plant communities (survey in Whittaker 1973) and are treated in some detail in the next paragraphs.

Microfungi, on the other hand, are the domain of soil microbiologists. Communities of microfungi cannot be studied directly in the field by inventories of sporocarps. All methods are based on isolation of fungi from certain substrates on artificial media in the laboratory. Many different techniques and culture media are in use (see chapter 7.3). Problems of these methods are (a) selectivity for particular ecological groups of fungi (e.g. Warcup 1951; Boois 1976); (b) the numbers of colonies on the plates are not necessarily representative of the different species in the soil; (c) part of the cultured species may in fact not belong to the active fungi in the sampled substrate, being only present as dormant spores (Eckblad 1978). The selectivity of these methods is clearly demonstrated by the almost complete absence of larger basidiomycetes from these isolates. Surveys of fungi isolated from e.g. oak forest soils (Witkamp 1960), coniferous litter (Brandsberg 1966), *Calluna* heathland soils (Sewell 1959a, b) and grasslands (Apinis 1958) do not list larger basidiomycetes although basidiocarps of litter decomposing species are abundant. Some special techniques have been developed to improve the results (e.g. Warcup and Talbot 1962), but they are not yet very successful.

The integration of methods for the analysis of macrofungal and microfungal communities, and the integration of the results is one of the important challenges for future research. For the time being the two approaches are both important and supplement each other.

2.3. Approaches to ecology and coenology of macrofungi

The entire field of ecological and coenological studies on fungi may be covered by the term *myco-ecology* (Cooke 1948). the following main directions are distinguished after Arnolds (1981):

a) The *mycocoenological* (or *mycosociological*) approach comprises quantitative inventories of all (macro)fungi in stands of well-defined plant communities or selected habitats, usually by means of repeated qualitative and quantitative analysis of selected plots. The first studies of this kind were published by Haas (1932), Höfler (1937), Leischner-Siska (1939) and Wilkins *et al.* (1937—1940). More recently important methodological contributions were made by e.g. Barkman (1976a, 1987), Arnolds, (1981, 1982), Thoen (1977) and Winterhoff (1975, 1984). Its methods and problems are extensively discussed in this chapter.

b) The *mycosynusial* approach resembles the mycocoenological approach in many respects, but instead of entire fungus communities only parts of them are investigated, either macrofungi in a special microhabitat (e.g. excrements, wood remains) or macrofungi belonging to a certain functional group (e.g. ectomycorrhizal fungi). Early studies of this kind were carried out by Moser (1949) on burnt sites and Pirk and Tüxen (1949) on excrements in pastures. Important contributions to its methodology were made by e.g. Darimont (1973) and Barkman (1970, 1973, 1976b). The synusial approach is extensively treated in this chapter.

c) The *mycofloristic-ecological* approach comprises studies of the mycoflora in more or less extensive areas, with a heterogeneous plant cover, paying special attention to preferences of macrofungi for certain plant communities or habitats. Such studies vary strongly concerning the variation of studied habitats, size of the area and the accuracy of the methods and results (Winterhoff, 1984). Some studies consist merely of an enumeration of species with superficial notes on their habitats in the studied area, others produce more quantitative data on habitat preferences of fungi, for instance lists of characteristic species of different plant communities. Well-known examples of the latter category are studies by Favre (1948, 1955, 1960), Einhellinger (1969, 1973, 1976, 1977), Kreisel (1957) and Winterhoff (1977). Advantages of this method are that rare or small (micro)habitats are automatically included and that an impression of habitat preference is attained much easier than with the laborious mycocoenological or mycosynusial approaches. On the other hand, the results are more subjective (e.g. by arbitrary distinction of plant communities in the field), not reproducable, and not quantitative. This method is not treated more extensively in this chapter. See also Winterhoff 1984.

d) The *phytocoenological* approach, including fungi, is carried out by vegetation scientists who pay accidental attention to the occurrence of macrofungi during their study of plant communities (e.g. Heinemann 1956; Carbiener *et al.* 1975). In view of the large temporal variation in sporocarp formation and the specialistic character of fungal taxonomy such contributions are of very limited value.

e) In the *autecological* approach one or a few related fungal taxa are studied in order to obtain information on their ecological range or preference for plant communities. Examples are studies by Arnolds (1974, 1980) on species of *Hygrocybe* and *Camarophyllus* in the Netherlands, Rammeloo and Van Hecke (1979) on the ecology of *Hirneola auricula-judae* (Bull.: Fr.) Berk. in Belgium, Andersson (1950) on 21 species of macrofungi in coastal dunes in Sweden and Bach (1956) on *Phaeolepiota aurea* (Mattuschka: Fr.) Maire ex Konr. and Maubl. The last-mentioned study includes experiments under controlled conditions. Many other experimental studies in the laboratory have been carried out with macrofungi from all functional groups, but these investigations clearly fall outside the scope of the present chapter.

f) In the *geographical* approach the occurrence of macrofungi is indicated on a detailed vegetation map, so that conclusions on their ecological preference can be drawn. It has only exceptionally been carried out, e.g. by Hendriks (1976) in an area with different types of juniper scrub in the Netherlands.

2.4. The status of fungal communities

2.4.1. *Notes on terminology*

Most myco-ecologists prefer to indicate concrete and abstract fungal communities with terms derived from well-known ecological terms, in particular borrowed from vegetation science, preceded by the prefix 'fungal' or 'myco-'. However, some authors claim that a special terminology is needed in view of the fundamental differences between green plants and fungi in taxonomic position and ecological strategies. The most elaborate terminology was proposed by Darimont (1973), who introduced for instance the French terms 'mycétation' (equivalent of fungus vegetation), 'fonge' (= mycoflora), 'mycotope' (= uniform habitat of a fungal community), 'synmycie' (= fungal society, see § 2.4.3), 'mycosynécie' (= combination of 'synmycie' and 'mycotope') and 'sociomycie' (= fungal synusia, see § 2.4.3). His terminology has hardly found support up to now, which is favourable for the accessibility of myco-ecological studies for a more general scientific audience.

2.4.2. *Fungal communities and mycocoenoses*

The term fungal community is used here in analogy to the term plant community, as a neutral term for any concrete assemblage of fungi that grows together in a certain uniform space, independent of its size and degree of heterogeneity in terms of habitat exploitation and substrate preference. An abstraction of such a unit may be indicated as fungal community type. Since a rather sharp methodological borderline exists between studies on communities of macro- and microfungi (§ 2.2) it is useful to distinguish *macrofungal*[1] and *microfungal* communities. It has to be stressed again that this separation is based on practical reasons only and therefore is completely artificial: it is not correlated with ecological, spatial, functional or taxonomic criteria (§ 2.2).

The term *mycocoenosis* is reserved here for the complete assemblage of fungi growing within a certain phytocoenosis and its environment or, in absence of green plants, within a uniform habitat (Smarda 1972, Arnolds 1981, 1988a). Arnolds (1988a) compared the properties of mycocoenoses and phytocoenoses in detail and concluded that a mycocoenosis is much more heterogeneous in ecological respect. An abstraction is a *mycocoenon* (analogous to phytocoenon). These terms have been currently used for communities of macrofungi only, which can be better called *macrofungocoenosis*[1] and *macrofungocoenon*, respectively (Arnolds 1988a). The term mycocoenosis has been used by some other authors (e.g. Kreisel 1957, Fellner 1987) in a more restricted meaning for interacting mycelia in a certain uniform substrate. This is similar to a mycosociety as described here (§ 2.4.3).

Macrofungocoenoses are usually studied in large plots with a variety of substrates and microhabitats (§ 2.5). This ecological heterogeneity is the most important disadvantage in comparison with the synusial approach (§ 2.4.3). However, this draw-back can partially be met by making accurate annotations on the microhabitats in which individual sporocarps or groups of sporocarps are observed (e.g. Barkman 1976b). The results are usually incorporated in phytocoenological classifications and rarely classified as units on their own, apparently due to their ecological heterogeneity (§ 2.10).

2.4.3. *Mycosocieties and synusiae*

Barkman (1973: 452) defined a society (abusively named synusia) as: "a structural part of a phytocoenosis inhabiting (a) a special microhabitat, with (b) a specific floristic composition and consisting of species that (c) belong to the same stratum and that do not differ fundamentally in either (d) periodicity or (e) way of exploitation of their environment". The abstract unit of societies is a synusia. These terms are in principle also appropriate for microcommunities of fungi (Barkman 1976b), named *mycosocieties* and

[1] From a linguistic point of view the terms *macromycete community* and *macromycocoenosis* would be preferable, but for practical reasons (e.g. resemblence of the latter to macromicrocoenosis) the term *macrofungocoenosis* if preferred.

mycosynusiae, respectively, or, when restricted to macrofungi, *macrofungosocieties* and *macrofungosynusiae.*

Some authors introduced special terms for these units in view of the special taxonomic and ecological position of fungi, e.g. 'synmycie' (mycosociety) and 'sociomycie' (mycosynusia) by Darimont (1973: 43) and 'consortium' (mycosynusia) by Kalamees (1965, 1979). On the other hand, Kreisel (1957) and Fellner (1987) called societies 'mycocoenoses' (cf. § 2.4.2).

Societies are usually investigated by means of small plots on a homogeneous substrate, in practice mainly on substrates with distinct spatial borders, such as burnt places, dung, stumps and trunks of trees. Macrofungosocieties on more diffuse substrates, such as litter and tree roots (mycorrhizal fungi), are rarely studied. Arnolds (1981, 1988a) amply discussed problems in the application of the synusial approach to macrofungi, which are summarized here:

a) The distinction of homogeneous microhabitats is often problematic, since many microhabitats consist of very small units (e.g. fallen leaves, twigs, fruits, herbaceous stems, excrements), offering space to only one or few species. The assessment of uniformity in these circumstances is very arbitrary.
b) The demand for a specific floristic composition can only be tested in the few cases of species-rich macrofungosocieties. For example, cones of *Pinus sylvestris* L. are inhabited in Western Europe by either *Auriscalpium vulgare* S. F. Gray or *Strobilurus tenacellus* (Pers:Fr.) Sing., *S. stephanocystis* (Hora) Sing. or *Baeospora myosura* (Fr.:Fr.) Sing. These species are hardly ever found together on a single cone. Consequently it is arbitrary to assign them to one synusia, based on assumed (only after our perception) homogeneity of substrate, or to four synusiae on the basis of different floristic composition.
c) On the basis of differences in the way of exploitation of the environment a distinction should be made between saprotrophic (saprophytic), biotrophic (symbiotic), mycorrhizal and necrotrophic (parasitic) macrofungosynusiae. This demand is stressed by Barkman (1973) and Fellner (1987). However, it is impossible to distinguish these functional (or trophic) groups with certainty in the field on the basis of characteristics of the sporocarps. Species belonging to different groups are often fruiting side by side in the same microhabitat, e.g. ectomycorrhizal and saprotrophic macrofungi in soil. Therefore, species have to be assigned to one of these groups, either by laborious tests in the laboratory (Schenck 1982) or by application of, very incomplete, literature accounts. For many important groups the way of habitat exploitation is not yet known, e.g. Clavariaceae, *Entoloma* and Hygrophoraceae.

2.4.4. *Ecological groups and niche-substrate groups*

An *ecological group* is a list of species with great resemblance in ecological optimum or range concerning certain relevant environmental factors. Differences between ecological groups on the one hand and societies (synusiae) and sociological groups on the other hand, were outlined by Barkman (1970: 106). An ecological group is an abstract list of species without spatial limits, whereas a society is a concrete community of individuals with spatial limits. A species can belong to several synusiae, but can be assigned to only one ecological group of a certain rank.

It has been argued that the synusiae, distinguished in myco-ecology, are in fact more similar to ecological groups (Arnolds 1981, 1988a, Barkman 1987) because in practice most macrofungosynusia are necessarily based on (supposed) uniformity of substrate and way of habitat exploitation and not on similarities in floristic composition (§ 2.5.2), in other words not on coenological but on ecological criteria (e.g. 'synusiae' described by Darimont (1973), Barkman (1976b) and Fellner (1987). An exception on this thesis are for instance the synusiae on tree stumps described by Runge (1980), which meet the criteria of true synusiae.

However, it is unsatisfactory that each fungal species should be assigned to only one ecological group since many species are able to grow on more than one substrate or have several possibilities to exploit the habitat. Properties of synusiae and ecological groups are combined in *niche-substrate groups* (Arnolds, 1988a) or *guilds* (Barkman, 1987), defined as lists of species growing on a similar substrate in a similar microhabitat and exploiting the environment in a similar way. The application of these groups is demonstrated by Arnolds (1981, 1988a) and in the chapters 5 and 6.

2.4.5. *Types of dependence in mycocoenoses and consequences for analysis and classification*

In view of their heterotrophic way of life all fungi are *ecologically dependent* on green plants either directly (e.g. mycorrhizal symbionts, saprotrophs on litter) or indirectly (e.g. fungi on dung, dead corpses and insects). Besides, ecological dependence can be unilateral (e.g. parasites) or mutual (e.g. mycorrhizal symbionts). Barkman (1973) distinguished next to ecological dependence two other types of dependence: *geographical dependence* (restricted to the same area as certain organisms or plant communities) and *syntaxonomic dependence* (restricted to a certain community type or syntaxon). All symbiotic fungi with direct ecological dependence are also dependent in geographical respect. This also applies to most saprotrophic fungi on plant remains, but the substrate may be moved by natural factors (wind, water) or human interference (e.g. transport of wood), so that the geographical dependence is lost. A number of examples were given by Arnolds (1988a). Fungi which are indirectly dependent in ecological respect are often not geographically dependent on a certain plant community since the substrate can be easily moved to different environments (e.g. dung, deposited in forests by animals grazing in pastures).

These considerations on dependence are important for the analytical and synthetical methods in mycocoenology. A distinction has to be made between *proper fungi*, which are ecologically and geographically dependent on organisms living in the investigated biocoenosis, and *alien fungi*, which are not dependent and therefore occur only accidentally in the studied biocoenosis (Arnolds 1981). The alien fungi should be either excluded from the interpretation of mycocoenological data or treated separately. It depends on the investigated communities which fungi have to be regarded as alien.

Two examples may elucidate the importance of these concepts. Rudnicka-Jezierska (1969) studied plots in a *Corynephorus* community and listed 15 species of macrofungi, including at least 5 obligate mycorrhizal symbionts of neighbouring trees, which clearly do not belong to the investigated communities and have to be regarded as alien. Comparison with data from *Corynephorus* communities without trees (Arnolds 1981) is therefore difficult. Pirk and Tüxen (1957) studied mycocoenological plots in dry heathlands (Calluno-Genistetum typicum). More than half of the fungi, mentioned by these authors, are mycorrhizal symbionts with trees. Among the 11 'probable character species' of the Calluno-Genistetum at least 7 are mycorrhizal with trees, which is in fact a contradiction in terminis. It is obvious that they should have been excluded as alien species. Other examples were given by Arnolds (1981, 1988a).

The easiest way to avoid such complications is to select plots without disturbing elements in the direct surroundings (Winterhoff, 1984). However, some types of alien fungi cannot be excluded from the plots in advance (e.g. coprophytic fungi), but only afterwards during processing of the data.

2.5. Selection of sample plots

2.5.1. *Criteria for sample plots*

Most mycocoenological work is based on the analysis of a small number of large plots, selected on the basis of criteria coming from phytocoenology, viz. uniformity (homogeneity) (§ 2.5.2), minimum area (§ 2.5.3) and representativity (§ 2.5.4). A rarely used, alternative method is the selection of a large number of small plots (e.g. 1 m^2) in a certain type of plant community (e.g. Lange 1948). Parker-Rhodes (1951, 1955) analysed the preference of macrofungi by means of random line transects. This approach was discussed and criticized by Arnolds (1981: 30).

2.5.2. *Uniformity of macrofungocoenoses*

A phytocoenosis is considered to be uniform (or 'homogeneous') when no obvious structural boundaries are visible in the stand or the habitat and the floristic composition is uniform, i.e. similar in different partial plots (Westhoff and Van der Maarel 1973: 632). Uniformity of phytocoenoses is usually judged subjectively. It is impossible to estimate or determine uniformity of macrofungocoenoses in this way since at any given moment only part of the species present is fruiting (see 2.8) and since the sporocarps do not constitute a distinct structural entity. Uniformity can be determined afterwards when a study has been finished, provided that sporocarps were mapped or that small partial plots were investigated. One of the classical methods to test uniformity is based on the frequencies of species in a series of plots of equal size, distributed in an assumed uniform phytocoenosis (Braun-Blanquet 1964). The distribution patterns of species over frequency classes are considered as an indication of the degree of uniformity. In a heterogeneous stand the proportion of species with high frequency is relatively low and the distribution over frequency classes is J-shaped, whereas in a uniform stand the proportion of species with high frequency is relatively high and the distribution over the classes is U-shaped. However, this distribution pattern is dependent on the plot size: in larger plots a larger proportion of species is common to all plots, irrespective of the degree of uniformity.

An analysis of this kind was carried out for macrofungi by Thoen (1977), who studied during one autumn the macrofungocoenoses in 5 nesting plots of increasing size within two, seemingly uniform, stands of *Picea abies* in Belgium (Figure 1). As Thoen (l.c.) pointed out, the results agree fundamentally with the results obtained in phytocoenoses: the number of species with high frequency increases with the size of the plots. However, the plot size in which the number of species in class 5 (80—100% of plots) surpasses those in class 1 (> 0—20% of plots) is considerably larger for macrofungi than for green plants. In other words: the investigated stands are uniform for both green plants and macrofungi, but the distribution patterns of fungi are on the average coarser and/or more irregular than those of green plants.

This result is confirmed by some other studies on distribution patterns of

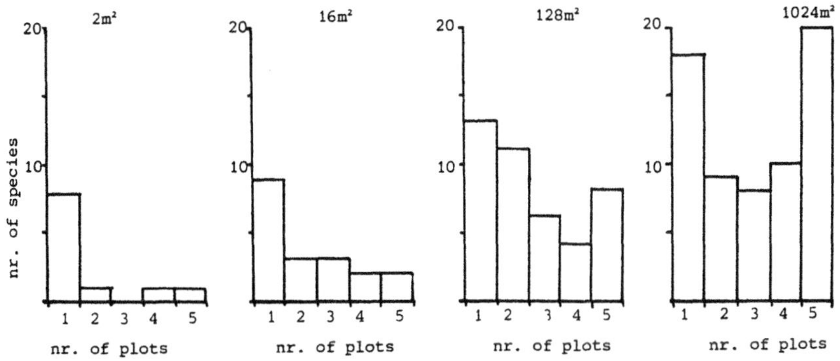

Fig. 1. Distribution of species of macrofungi over frequency classes (in this case number of plots) in plots of different size in a uniform stand of *Picea abies* in East Belgium. Within the stand 5 nesting plots of increasing size (up to 1024 m^2) were studied (after Thoen 1977).

macrofungi in forest stands. It seems that within a functional group of macrofungi species tend to exclude each other and usually occur less intricately intermixed than green plants. For example, Murakami (1987) demonstrated considerable spatial exclusion of 5 Russula species in a homogeneous deciduous forest in Japan (Figure 2). De Vries (1990) found that lignicolous fungi in a homogeneous plantation of *Picea* were unevenly distributed over partial plots: In 4 plots of 8 m^2 only 2 species were present in all 4 plots, 5 species were observed in 3 plots, 5 species in 2 plots and 13 species in only 1 plot. Villeneuve *et al.* (1989) studied macrofungocoenoses in 5 replicate plots of 400 m^2 in 4 forest associations near Québec (Canada). The frequency of fungi was determined in one hundred microquadrats of 2 × 2 m each at 7 to 10 day intervals during two years. Among the 195 recorded species only 2 had a frequency over 20% in one of the associations and 9 species had a frequency between 10 and 20% in one or more associations.

Since many species of macrofungi are associated with green plants, these species may occur clustered around these plants. Examples of clustered fruiting of macrofungi were given by Jansen (1974, Figure 3). In this example the occurrence of some fungal species is positively correlated with the presence of moss-rich patches within the phytocoenosis, either as a result of direct ecological relations with bryophytes (*Galerina decipiens* Smith & Sing. = *G. hypnorum* s.l.), or due to indirect factors, such as the absence of a thick litter layer (*Inocybe xanthomelas* Kühn. & Bours., *Cordyceps ophioglossiodes* (Ehrh.: Fr.) Link). In phytocoenology such concentrations of species may be regarded as separate phytocoenoses (Westhoff and Van der Maarel 1973:

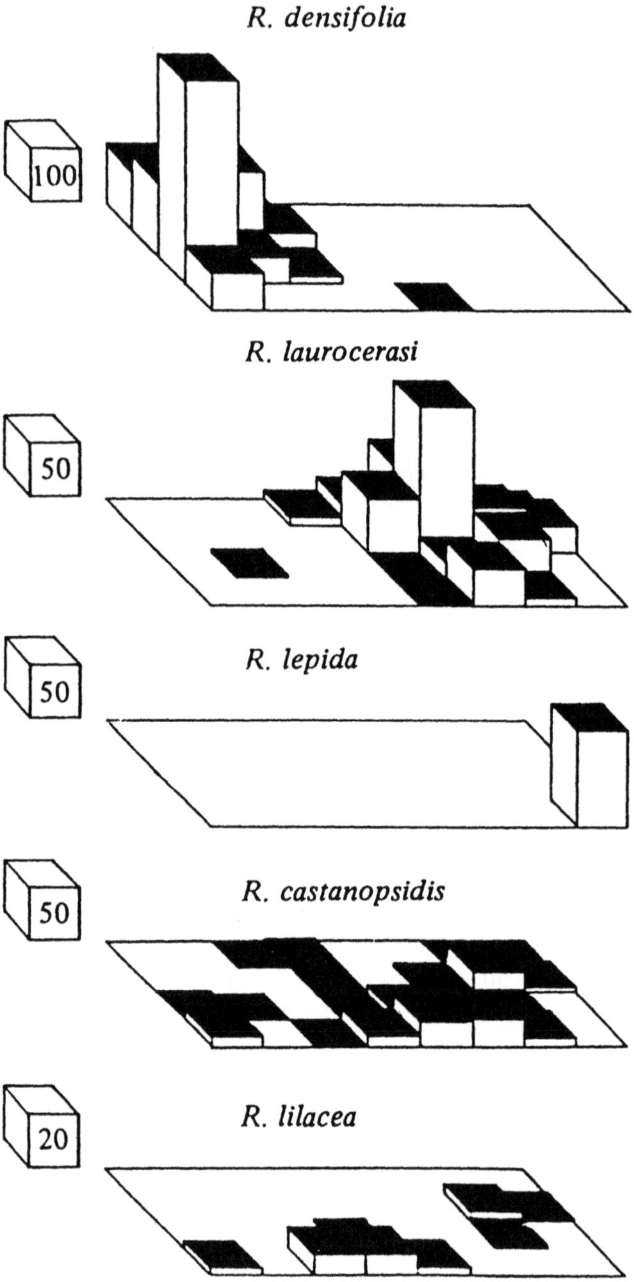

Fig. 2. Spatial distribution of 5 *Russula* species in 32 sub-quadrats (5 × 5 m) in a forest of *Castanopsis cuspidata* and *Pasania edulis* on Mt. Tachibana, Fukuoka, Japan. The numbers to the left indicate the abundance of basidiocarps, over a period of two years (after Murakami 1987).

Abundance of fungi is indicated as follows:

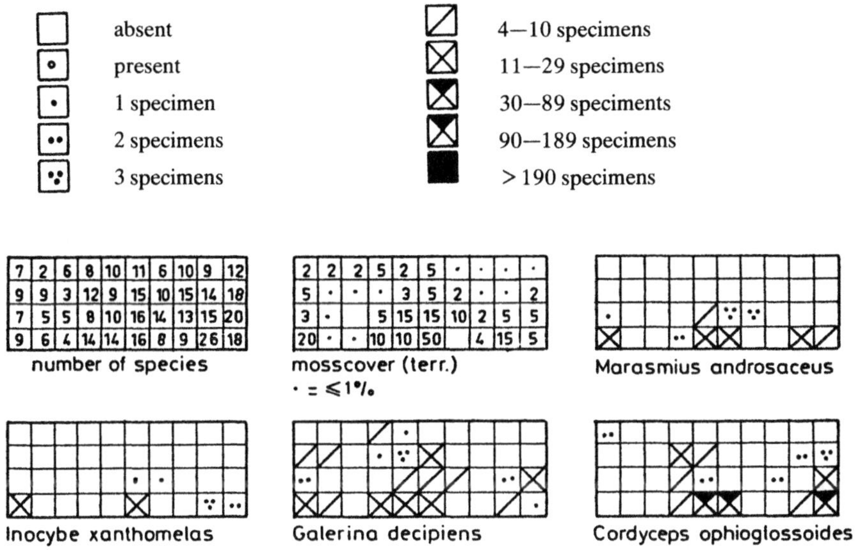

Fig. 3. Distribution of number of species of macrofungi, of moss cover (proportion of subplots), and distribution and abundance of some fungi in 40 subplots of 25 m^2 each in a stand dominated by *Quercus robur* on dry, poor sand (Dicrano-Quercetum) in Drenthe, the Netherlands (from Jansen 1984).

633), but a rigid application of this criterion on macrofungocoenoses would lead to an unwanted fragmentation of plots.

In view of the mentioned practical difficulties, mycocoenological plots are nearly always selected on the basis of uniformity of the phytocoenosis (or the environment if green plants are missing). Even more problematic is the use of the uniformity criterion in the selection of plots for macrofungal societies (§ 2.4.3). The estimation of uniformity is only possible for microhabitats with a surface large enough to accommodate various species of macrofungi, e.g. a litter layer in a forest, tree trunks and stumps, burnt sites, dung heaps. However, most microhabitats for macrofungi consist of very small units, e.g. culms, twigs, branches and animal excrements. In practice the plot selection is not based on the floristic composition of plants or fungi, but on hypothetical uniformity of the substrate, usually within a certain type of phytocoenosis. This problem is more amply discussed by Arnolds (1988a).

The frequent application of criteria, not based on composition and structure of the macrofungocoenoses themselves, limits the possibilities of an independent classification of these units (see § 2.10).

2.5.3. *Size of mycocoenological plots*

Mycocoenological plots should meet the criteria for the analytic minimal (or representative) area, formulated at present in phytocoenology as an adequate sample of species of regular occurrence in the stand (Westhoff and Van der Maarel 1973: 636). It is extremely difficult to define the minimal area concept on a mathematical basis. The traditional minimal area curves (relation between plot size and number of species) for phytocoenoses appear to be unsatisfactory, since in fact the number of species usually remains increasing with a larger plot size (e.g. Barkman 1984).

The determination of minimal areas for macrofungocoenoses is even more difficult since the periodicity and fluctuations of sporocarps (§ 2.8) require in fact repeated analysis of each plot size during several years. The attempts made so far do usually not acknowledge this complication. Winterhoff (1975) constructed a species-area curve for a dune grassland in Germany, based on unique sampling during peak productivity. He found on 100 m^2 only 25% and on 1000 m^2 50% of the species present in the entire stand of 6700 m^2. The representative area of the plant community was in order of a few m^2. Arnolds (1981) found the same picture when comparing grass-heathland plots of different size, studied during three successive years. The number of macrofungal species was still increasing at 400 m^2. De Vries (1988) determined the relation between plot surface and number of species of wood inhabiting macrofungi in a homogeneous *Picea* plantation. He did not even find a reduced increase of the species number with the largest doubling of the plot from 500 to 1000 m^2. The most extensive study of the relation between numbers of macrofungal species and plot size was carried out by Thoen (1977) in two mature uniform plantations of *Picea abies* (L.) Karsten in East-Belgium (Figure 4). In each stand he made an inventory of macrofungi in 5 sets of nesting plots with a size increasing from 1 to 1024 m^2 in September and October 1975. His results confirm the observations by other authors that the number of species remains increasing at the last doublings of the surface. Thoen (l.c.) pointed out that the increase per added m^2 is becoming strongly reduced with increasing plot size, however. The two stands show comparable minimal area curves (Figure 4). Thoen compared the curves for macrofungi with curves for green plants (phanerogams and bryophytes) in the same sets of plots. The general appearance of these curves is comparable, but the increase of fungal species is steeper: at a size of 16 m^2 on the average half of the number of green plants at 1024 m^2 is present, but only 30% of the number of macrofungi. The number of macrofungi is lower than the number of green plants up to a surface of 4 m^2, but higher from a surface of 8 m^2 onwards.

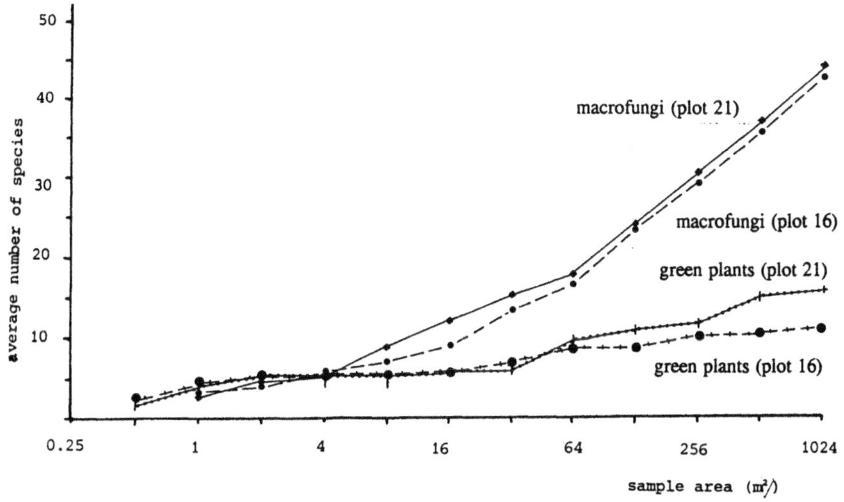

Fig. 4. The relations between the average number of species of macrofungi and green plants (phanerogams and bryophytes), and the plot size (5 replicate plots between 0.25 and 1024 m^2 of each size) in two stand of *Picea abies* in Belgium (after data by Thoen 1977). On the X-axis the plot size is indicated in a logarithmic scale.

In a second experimental design Thoen (1977) investigated during one year the macrofungocoenoses in 15 or 16 plots, 450 m^2 each, within two mature *Picea* stands in East Belgium. One stand was situated on moist soil and characterized by various *Sphagnum* species; the other stand was drier and rich in the moss *Leucobryum glaucum* (Hedw.) Angstr. On the basis of these plots he was able to construct a traditional minimal area curve from plots of 450 m^2 (based on 15 and 16 plots, respectively) up to 7200 (6750) m^2 (based on one composite plot). In the two stands a continuous increase of the number of species is observed (Figure 5), but the increase in the dry plot is stronger. Consequently it seems that a minimal area in strictly theoretical sense does neither exist for green plants, nor for macrofungi. However, it is useful to distinguish a, necessarily arbitrary, representative sample area. All available data indicate that this area is considerably larger for macrofungocoenoses than for phytocoenoses in both forests and grassland communities. According to Winterhoff (1984) this is caused by the considerable size of many fungal mycelia (e.g. fairy rings may reach a diameter of over 50 metres), the clustered occurrence of many species, the small proportion of a phytocoenosis that is available for the numerous substrate specialists and the rarity of many species. It may be added that mycelia are more inclined to mutual exclusion (§ 2.5.2).

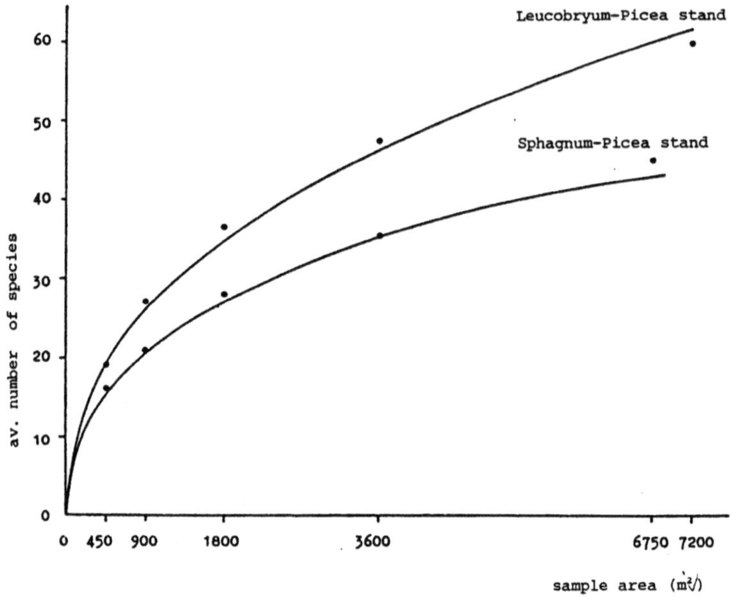

Fig. 5. The relation between the average number of species of macrofungi and the plot size (from 450 m^2 onwards) in two stands of *Picea abies* in Belgium, based on the analysis of 15 and 16 plots of 450 m^2 each in a *Sphagnum*-rich and *Leucobryum*-rich stand, respectively (after Thoen 1977).

As a practical compromise between (1) attempted completeness, (2) availability of uniform plots and (3) necessary efforts, most mycocoenologists agree now in a plot size of ± 1000 m^2 in scrubs and forests and ± 500 m^2 in grass- and heathlands (Winterhoff 1984, Barkman 1987).

2.5.4. *Representitivity and number of plots*

Westhoff and Van der Maarel (1973: 631) formulated the demand for representativity of phytocoenological plots as follows: "A selection of relevés is desired that will effectively represent the variation in the vegetation under study, the samples being so chosen that they will not represent different phytocoena disproportionately and will not include mixed, incomplete, or unstable stands." The situation for mycocoenological plots is identical, but again the judgement of representativity depends in practice mainly on characteristics of the plant community. The factors determining representativity are not necessarily similar for green plants and macrofungi, but this is difficult to assess in advance. An indication for this supposition is that Arnolds (1981) found in sets of plots in certain grassland associations much more variability in the number of species of macrofungi than of green plants, for instance in the Ornithopodo-Corynephoretum 47—53 species of green plants and 13—28 proper species of macrofungi per plot; in the Festuco-Thymetum 59—81 and 24—55; in the Agrostietum tenuis 46—55 and 34—62 species; and in the Cynosuro-Lolietum 34—58 and 2—43 species, respectively.

An important factor in representativity of mycocoenological plots is the comparability of variation in microhabitats. For instance, in forests the abundance and variation of woody substrate will determine to a large extent the number of lignicolous fungi. When quantities of deadwood are very different in fact a correction factor should be applied for the calculation of abundance of lignicolous fungi (e.g. Jansen 1984; Barkman 1987). Similarly, the variation in tree species in a plot will determine to a large extent the variation in mycorrhizal fungi to be expected.

The number of plots should preferably be 10 or more to represent the variation in a certain phytocoenon or mycocoenon. Because of the large efforts needed for mycocoenological analyses (§ 2.8) such numbers are rarely realized (cf. Winterhoff 1984: 235). From a practical point of view a minimum set of 5 plots per mycocoenon is desirable. For the analysis of small-scale macrofungosocieties much larger numbers of plots are required in order to obtain reliable results (examples in Winterhoff 1984: 235).

2.6. Qualitative analysis of macrofungocoenoses

The qualitative analysis of a macrofungocoenosis is nothing else than the compilation of a list of species of macrofungi occurring in a plot. This is fundamentally a simple procedure, but in practice some complications exist:

a) Identification of many taxa is difficult and demands specialized literature. In order to facilitate future revisions and the application of new taxonomic concepts it is necessary to preserve voucher specimens, at least of critical taxa and by preference in a public herbarium. The preservation of all taxa is strongly advocated since taxonomic concepts are still in development. It is also recommended to publish taxonomic notes on critical and rare taxa. This is a good habit in the mycofloristical-ecological approach (§ 2.3; e.g. Favre 1948, 1955, 1960, Einhellinger 1969, 1973), but has been applied less often in mycocoenological studies (e.g. Arnolds 1981, Jansen 1984). Unidentified species can be indicated with provisional names or numbers (e.g. Cooke 1955).

b) Many taxa are known in different concepts, e.g. *Hygrocybe conica*

(Schaeff.: Fr.) Kummer sensu stricto and sensu lato (incl. *H. conicoides* (P.D. Orton) Orton & Watl., *H. conicopalustris* R. Haller, *H. nigrescens* (Quél.) Kühner, *H. tristis* (Pers.) Moell., *H. olivaceonigra* (P.D. Orton) Moser and *H. riparia* Kreisel); *Cortinarius rigidus* Fr. sensu J. Lange and sensu Kühn. & Romagn. In addition nomenclature is in many groups confusing. Therefore it is necessary to indicate the consulted literature and nomenclatural basis.

c) The circumscription of the group of macrofungi is variable (§ 2.2, Table I). Different definitions may have important consequences for comparison of characteristics (e.g. species numbers) in different macrofungocoena. Consequently the studied taxon groups should always be explicitly stated.

d) Some authors neglect unmistakable macrofungi (e.g. some agarics) in their plots because of their small size, hidden way of life and/or taxonomic problems (e.g. the genus *Galerina*). This practice should be avoided. If necessary, collective species can be distinguished.

2.7. Quantitative analysis of macrofungocoenoses

The quantitative analysis of a macrofungocoenosis involves the expression of the quantity of each species, and is necessarily based on the presence of sporocarps. This is a disadvantage because it is uncertain which property of sporocarps correlates best with mycelium extension or activity in the substrate, which are in fact the parameters with highest ecological significance. Most authors use the individual sporocarps as units for quantitative analysis (e.g. Bohus and Babos 1960, Barkman 1976a, 1987, Sadowska 1973, Thoen 1977, Arnolds 1981, Jansen 1984). It is in most cases an unequivocal character, with the exception of species with aggregated or confluent sporocarps, especially resupinate Aphyllophorales. Some mycocoenologists (e.g. Haas 1932, Parker-Rhodes 1951, Darimont 1973) prefer the distinction of groups of sporocarps in the field ('station' according to Darimont, 'locus' according to Parker-Rhodes, l.c.), which are considered to be a better expression of the number of mycelia. Thoen (1977) and Arnolds (1981: 114) formulated fundamental and practical objections against the latter approach, the most important being the doubtful basis of the above supposition, the subjectivity of the distinction of sporocarp groups and difficulties in application of this concept in relation to the dynamics of fruiting.

For the expression of the quantity of sporocarps various criteria are used:

a) *Abundance* or *density*: the actual number of sporocarps in a plot, in the case of density calculated for a standard-surface (often 1000 m^2). Abundance can be established by counting or estimating. Sporocarp counts are in principle more exact, offer all possibilities for quantitative processing of data, but are very laborious (examples: Bohus and Babos 1960, Smarda 1968, Sadowska 1973, Arnolds 1981). Estimation of abundance is more often used and is sufficient for comparison and classification of mycocoenoses. However, possibilities for data processing are limited. For example, it is impossible to derive productivity figures in an indirect way (see b). Estimation of abundance is always carried out with the aid of a scale with increasing intervals to the higher values. Examples are given in Table II. For a comparison of various scales, see Arnolds (1981: 117).

b) *Weight* and *size*: Höfler (1937) and Bohus and Babos (1960) considered the mass of sporocarps as the most representative expression of the activity of mycelia and used it as quantitative measure. Their assumption has not been proven, but variation in sporocarp weight is considerable indeed. For example, Arnolds (1981) found in grassland agarics average sporocarp weights ranging from 0.05 mgr (in *Mycena pudica* Hora) to 13 gr (in *Macrolepiota procera* (Scop.: Fr.) Sing.), consequently differing a factor 260 000. It is obvious that the latter species must have a larger mycelium volume per sporocarp than the former. On the other hand, in phanerogams no direct relation exists between size of the plant and weight of fruits. Since sporocarp weight is a relatively constant character of each species it is not of direct importance for the diagnostic value when comparing different mycocoenoses, but the measurement of sporocarp productivity is a useful aim in itself (e.g. Ohenoja 1978, 1983). Productivity can be determined in a direct way by harvesting all sporocarps and weighing them fresh and/or dried, or in an indirect way by counting them and multiplying the abundance by the average (dry) weight of a representative sample of mature sporocarps (Arnolds 1981, Senn-Irlet 1986). The latter is less accurate, but also less laborious and less destructive. Moreover, it has the advantage that underestimation of the productivity by reduced weight of young or overmature sporocarps is prevented.

c) *Sociability*: a measure for the spatial arrangement of sporocarps in groups, ranging from isolated to large clusters. Sociability is usually estimated, using the scale proposed by Haas (1932) or related scales (Darimont 1973). Kalamees (1968) regarded sociability as much more important than abundance. This importance is only stressed by authors who determine abundance of sporocarp groups. A combination with sociability is than a rough indication of sporocarp abundance. When the latter property is directly determined, sociability is, like in phanerogams, an additional structural character which is more or less constant for each

Table II. Example of scales for the estimation of abundance (density) of macrofungi in mycocoenological plots (from Arnolds 1981)

A. Scales for the estimation of abundance (density) of mycelia

References:

Haas 1932		Darimont 1973		Winterhoff 1975	
+	1 group	RR	1 group	+	1 group
1	very few gr.	R	1—3 gr.	1	2—5 gr.
2	very scattered	AR-AC	4—10 gr.	2	6—10 gr.
3	Scattered	C	11—25 gr.	3	scattered
4	many gr.	CC	26—100 gr.	4	many gr.
5	in masses	CCC	> 100 gr.	5	everywhere
			per ha.		on 100 m²

B. Scales for the estimation of abundance (density) of sporocarps

References:

Höfler 1937	Pirk 1948	Moser 1949	Guminska 1966	Barkman 1976	Arnolds 1981
+ one or a few	1	1	1—5	1—2	
1 sparse	2—5	2—5	6—10	3—10	1—3
2 not numerous	6—10	6—50	11—50	10—100	3—10
3 fairly abundant	11—20	51—100	51—100	100—500	10—30
4 abundant	21—50	101—500	101—500	> 500	30—100
5 very abundant	> 50	> 500	> 500		100—300
6					300—1000
7					1000—3000
8					3000—10000
9					> 100000
	on 100 m²	in a stand	on 1000 m²	on 1000 m²	on 1000 m²

species and has only limited diagnostic significance (Westhoff and Van der Maarel 1973: 641).

d) *Spatial frequency*: the number of (partial) plots in which a species is present. As a quantitative measure of species performance the use of frequency is almost restricted to the synusial approach, working with large numbers of small, spatially separated plots, e.g. by Petersen (1970) for fungi on burnt sites and by Lange (1948) in his study of 1 m^2 quadrats in a peat bog. Jansen (1984) determined frequency by dividing a 1000 m^2 plot into 40 squares of 25 m^2 each. The results proved to be of little diagnostic value, but were interesting to demonstrate spatial patterns of some fungal species within the plots (Figure 3).

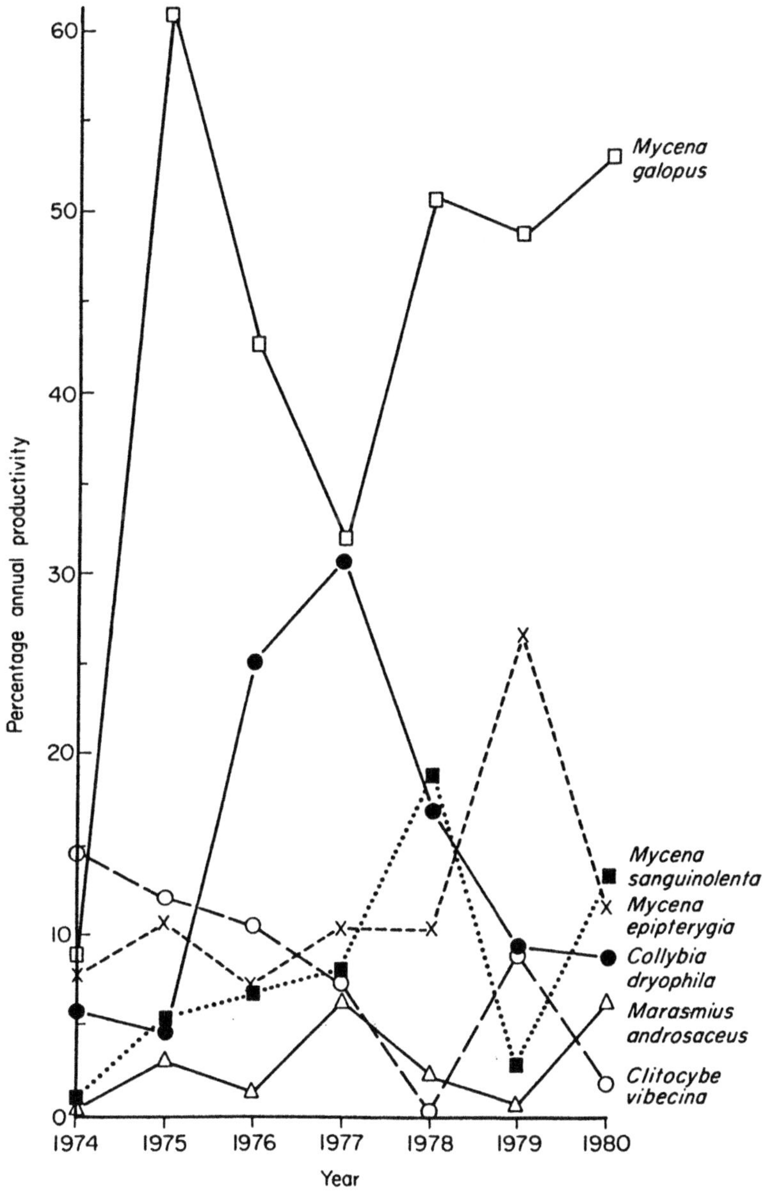

Fig. 6. Changes in carpophore productivity of six dominant, saprophytic species of macrofungi in a stand of *Erica tetralix* (500 m^2) in Drenthe, the Netherlands, in a period of 7 years, expressed as the percentage of the total carpophore productivity in each year (after Arnolds 1988b).

2.8. Frequency and duration of mycocoenological analysis

2.8.1. *Types of temporal variation in sporocarp production*

All mycocoenologists agree that it is necessary to sample mycocoenological plots more often than once in order to obtain a reliable survey of its mycofloristic composition and the abundance of sporocarps. This is caused by three factors (cf. Arnolds 1985, 1988b).

a) Short *duration* of sporocarps. Most sporocarps of macrofungi live a few days (e.g. *Mycena* spp., *Galerina* spp.) to weeks (e.g. boletes, *Russula* spp., *Collybia* spp.) (e.g. Lange 1948, Guminska 1962). Richardson (1970) found in a Scottish pine plantation an average life span of sporocarps of 5 days in e.g. *Entoloma cetratum* (Fr.) Mos. and *Russula emetica* Fr., up to 16 days in *Hygrophoropsis aurantiaca* (Wulf.: Fr.) Maire. It means that in these circumstances 50% of the sporocarps will be missed when sampling is done once a fortnight. Sporocarps of small coprophilous *Coprinus* species last only a few hours, but of many resupinate Aphyllophorales, polypores and Gasteromycetes between 1 and 4 months. Perennial sporocarps are only known from a few polypores, e.g. *Fomes*, *Ganoderma*, *Phellinus* and *Oxyporus* (e.g. Jahn 1979). The duration of mycelia is much longer and is mainly limited by the available substrate and presence of competing mycelia. Examples are known of fairy rings up to 750 years old (Shantz and Piemeisel 1917, Kreisel and Ritter 1985).
b) *Periodicity* of sporocarps. At least in temperate regions most species of macrofungi have a distinct periodicity, producing sporocarps only during a certain period of the year. The periodicity is a specific character and only influenced by weather conditions to a limited extent. Guminska (1962) distinguished several types of fruiting rhythms, dependent on the length of the fruiting period and the moment of peak productivity within this period. In north-temperate regions most species are fruiting in late summer and autumn and a minority in spring, early summer or winter (see example in chapter 5, Figure 7). In mediterranean areas winter and spring are much more important fruiting periods. The combinations of species, fruiting in a certain period in a certain plant community, are known as *fungus aspects* or *seasonal aspects* and may be indicated by characteristic species (Höfler 1954). Considerable regional differences may exist: *Hygrocybe* species are fruiting in Northwestern Europe mainly in late autumn (October—November), but in the Northeastern United States mainly in summer (July—early September) (Arnolds unpubl. obs.). Species may also have a different periodicity in different habitats (e.g.

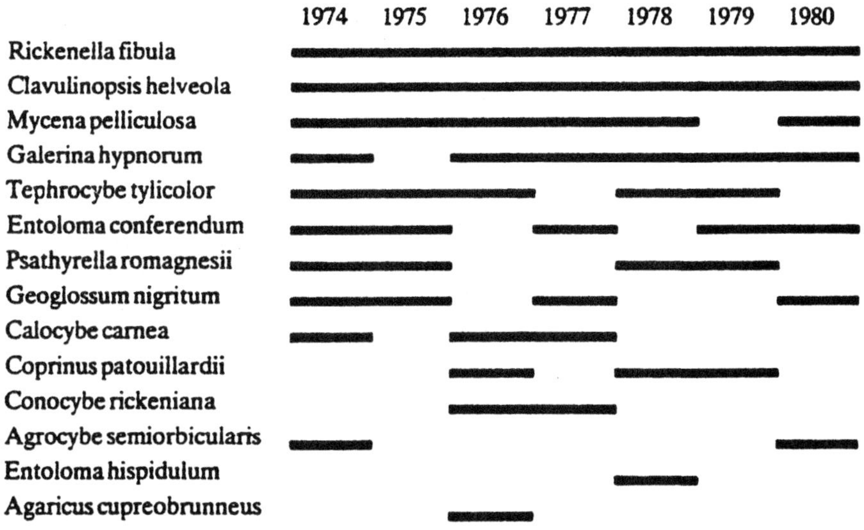

Fig. 7. Qualitative fluctuations of 14 selected species of macrofungi in a plot (400 m^2) in a poor, sandy grassland (Agrostietum tenuis) in Drenthe, the Netherlands, in the period 1974—1980 (after Arnolds 1981).

Barkman 1964, Einhellinger 1969, Arnolds 1982). In general fruiting is earlier in habitats with a high moisture content of the soil and/or air.

c) *Fluctuations* in fruiting, i.e. variations in numbers of sporocarps from one year to another. Fluctuations are quantitative when a species is present during all years considered and only the abundance of sporocarps is varying. Such fluctuations may amount to a factor 10—100 in successive years (Table III). Quantitative fluctuations have to a certain degree a general character because of a synchronical course of abundance figures for all or most species within a community. This phenomenon leads to the characterization of some years as 'good' or 'bad' mushroom years, for instance in Table III 1977 (all species with sporocarp density above average) and 1975 (all species below average), respectively. On the other hand, quantitative fluctuations may be more or less specific, as demonstrated in Figure 6. In this case the proportional contribution to the productivity is shown for 6 species. When these species should have synchronical fluctuations the curves would be theoretically horizontal lines. In fact fluctuations appear to be very considerable, in order of a factor 5—20.

Fluctuations are qualitative when sporocarps of a species are completely absent in a certain stand during one or more years (Figure 7).

Table III. Changes in abundance of 6 dominant species of macrofungi in a stand of *Erica tetralix* in Drenthe, the Netherlands, during the period 1974—1980, expressed as the total annual carpophore density per 1000 m^2 (from Arnolds in Acta Bot. Neerl. 37, 1988)

Species/Year	1974	1975	1976	1977	1978	1979	1980	Total	Average
Mycena sanguinolenta	132	342	1598	4486	11428	298	8432	26716	3817
Mycena galopus	282	868	2050	3956	6262	1132	5414	19964	2852
Marasmius androsaceus	118	302	562	5818	2206	146	5078	14230	2033
Mycena epipterygia	212	132	294	1048	1108	540	1038	4372	625
Collybia dryophila	26	8	164	488	284	30	122	1122	160
Clitocybe vibecina	122	46	134	226	28	58	54	641	92

Usually only a small minority of species in mycocoenological plots is found to be fruiting each year within a period of 3—5 years. Some cases are known of species fruiting in exactly the same locality with intervals of 10 years or more (e.g. Reijnders 1968, Reid 1974). Qualitative fluctuations logically lead to variations in species numbers in plots over the years (Figure 8). Fluctuations are mainly caused by weather conditions a short time before and during the potential fruiting period (e.g. Wilkins *et al.* 1937, Thoen 1976, 1977), but the mechanisms are not fully understood. Apparently also meteorological conditions over a longer period and other factors (litter productivity; vitality of trees in mycorrhizal fungi) may be important (Arnolds 1981, 1988b). In general warm, humid weather is favourable for rich fruiting, whereas severe frost (<-5 °C) and long drought are harmful to most species.

2.8.2. *Consequences for mycocoenological methods*

The complete sampling of all terrestrial sporocarps in a plot demands sampling with intervals of three to four days (Richardson 1970), but such an intensive program can be realized only in special cases. Consequently a practical compromise has to be found. The degree of completeness of mycocoenological sampling depends on three variables: plot size, frequency of observations within a year and duration of observations in years. Winterhoff (1984) calculated from the results of five mycocoenological plots in deciduous forests and grasslands, investigated during 4—5 years, that the minimum number of species per year ranged from 7 to 62% of the total number, the average number from 32 to 73% and the maximum number from 67 to 88%. The course of the cumulative number of species and the

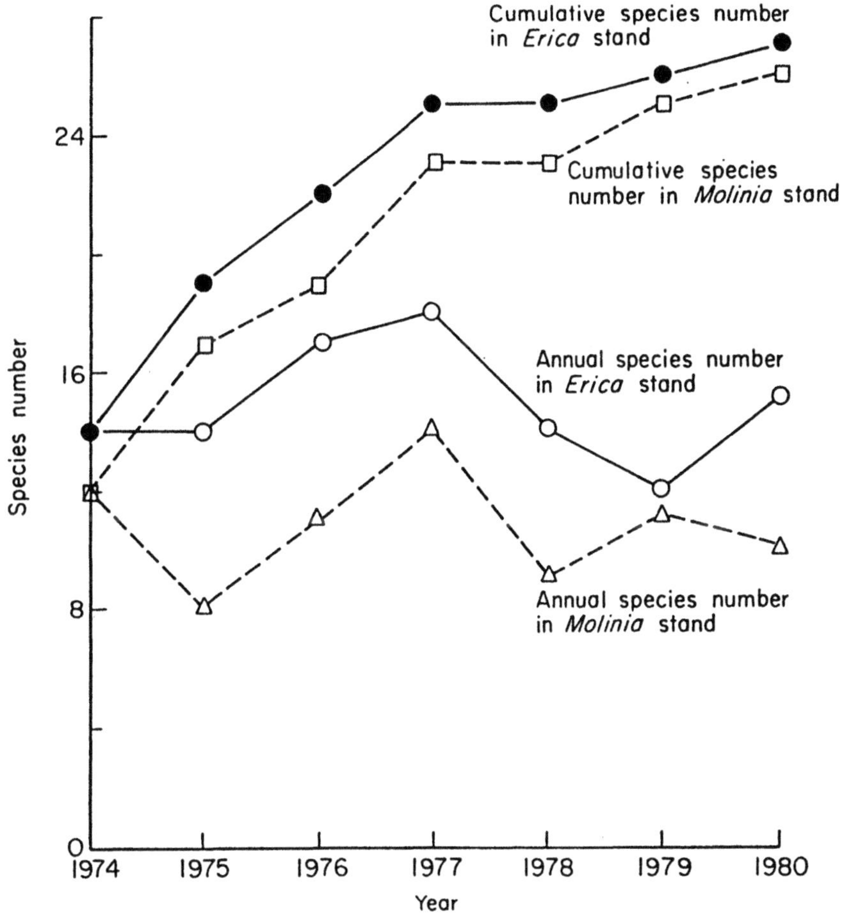

Fig. 8. Cumulative numbers of species and annual numbers of species of macrofungi fruiting in a stand of *Erica tetralix* and *Molinia caerulea* in Drenthe, the Netherlands, in the period 1974—1980 (after Arnolds 1988b).

annual numbers in a heathland plot are given in Figure 8. Arnolds (1981) calculated the percentage of species which can be expected in grassland plots with different sampling programs (Table IV). These indications can be reasonably used for other types of habitats as well. In my opinion it should be attempted to reach at least 75% of the total species number after 6 years, consequently for plots of 500—1000 m^2 sampling during 3 years each fortnight, 4 years each month or 5 years once in ±8 weeks (in the main season). Such sampling programs produce also representative figures on the

Table IV. Estimation of species number of macrofungi in grasslands, following different sampling procedures, expressed in the expected percentage of the total number of species after sampling each fortnight during six successive years (= 100%) (after Arnolds 1981)

	Duration in years					
	1	2	3	4	5	6
Frequency of sampling						
1 × per two weeks	50	70	85	92	97	100
1 × per four weeks	40	55	70	80	87	94
1 × per eight weeks (in the autumn)	28	45	60	72	80	87

abundance of the present macrofungi. Consequently, mycocoenological research is not only dependent on a representative plot size (§ 2.5.3), but also on a representative sampling procedure.

A complication which may be encountered during long-term sampling is the simultaneous succession of vegetation and environment. This factor is usually not regarded as important in (stable) forest communities, except when changes in external environment factors are very strong, as may be the case with changes in hydrology (drainage) and with high deposition of air pollutants (e.g. IJpelaar 1985). In more dynamic plant communities the results of mycocoenological studies may very well be influenced by successional trends (e.g. Arnolds 1988b). On some special substrates succession of fungi is even very rapid, in the order of months (dung, burnt sites) or years (tree stumps) (Kreisel and Dörfelt 1985).

2.9. Synthesis of mycocoenological data

2.9.1. *Synthesis of mycocoenological data for comparison between plots*

In phytocoenology the abundance of plant species in a plot is usually established only once and these data are the basis for comparisons between plots in the synthetical research phase (Braun-Blanquet 1964, Westhoff and Van der Maarel 1973: 643). In mycocoenology plots are usually analysed several times during several years (see § 2.8) and consequently an intermediate step is necessary, namely the calculation of one or more synthetical measures of the species quantity for each plot. In this respect many possibilities exist and they have, unfortunately, been used indeed, because the variety of parameters is sincerely hampering the possibilities for comparison of results of mycocoenological studies. Critical reviews on this subject and examples of calculations of synthetical measures were published by e.g. Barkman (1976a, 1987), Winterhoff (1984) and Arnolds (1981). Only the more important synthetical measures, with the proposed abbreviated indications, are mentioned here:

a) Maximum density of sporocarps during a single visit (mDCv according to Arnolds 1981; = absolute maximum abundance of carpophores (AMAC) according to Barkman 1976a). This expression is regarded to approach the potential fruiting capacity of a species. Its most important advantages above most other measures are that it can be calculated for each study, irrespective of the frequency of visits or duration in years, and that it is less influenced by these variables. It has been used by e.g. Haas (1932, for mycelia instead of sporocarps), Leischner-Siska (1939), Rudnicka-Jezierska (1969), Thoen (1970, 1971), Darimont (1973), Arnolds (1981), Jansen (1984).
b) Total density of sporocarps during all years (tDC) was propagated by Smarda (1968). It is a good parameter for comparison of plots studied by one author with a single method, but not appropriate for comparison with other methods since the results depend on duration and intensity of the research. A variation on this parameter was introduced by Bohus and Babos (1960) who calculated for each species the percentage of the total sporocarp number of all species in a plot, and called this 'dominance'. In this way the influence of intensity and duration of research is minimalised, indeed, but it is not a good importance value either since the result depends on the total sporocarp density. In other words, a species with a particular number of sporocarps may be 'dominant' in one plot, but with the same abundance it may be unimportant in another plot. A derived parameter is the total dry weight of sporocarps, for instance used by Senn-Irlet (1987; Table V).
c) The average annual density of sporocarps (aDCy) is a better expression since it is less dependent on the duration of research, but still on the frequency of visits. It was used as additional character by Arnolds (1981), Senn-Irlet (1987) and in slightly modified form by Guminska (1976).
d) The average density of sporocarps per visit (aDVc) seems to avoid the above disadvantages, but instead is very much dependent on the spreading of visits over the year. It makes much difference whether visits are concentrated in autumn or made at equal intervals throughout the year.
e) The total temporal frequency (tTFC) is the number of visits on which a species was observed during a mycocoenological analysis. It was used by e.g. Nespiak (1971) and Lisiewska (1978). It is not suited as importance value since it is dependent on duration of research, frequency and

Table V. Mycocoenological relévés during 3 successive years in a plot (98 m^2) in an alpine marsh community (Caricetum canescenti-fuscae) in Oberaar, Switzerland (alt. 2315 m) (after Senn-Irlet 1987, slightly modified)

Year	83	83	83	83	84	84	85	85	85	83	84	85		DWC
Month	7	8	8	9	8	8	7	8	9	DC	DC	DC	aDC	mg
Day	27	11	24	23	3	20	27	22	9	on	100	m^2		3y
Number of species	1	1	5	9	1	4	2	11	6	11	4	4	6.3	
Number of sporocarps	3	13	44	186	9	48	17	92	46	251	58	158	156	9695
Species, associated with bryophytes (parasites and saprophytes):														
Arrhenia lobata	3									3			1	33
Rickenella mellea		13	35	123		6	15	38	14	174	6	68	82.6	992
Galerina vittaeformis 2sp			1		9	17			7	1	27	7	11.5	350
Hypholoma myosotis				2					1	2		1	1.3	159
Galerina stagnina				2				1		2			1.3	117
Galerina subclavata				34						35			11.6	315
Galerina atkinsoniana				14						14			4.6	252
Rickenella fibula						14	2		8		14	10	8	96
Scutellinia aff. *mirabilis*								3				3	1.3	9
Ectomycorrhizal species:														
Laccaria laccata var. *tatrensis*			3	8		11		26	11	11	11	38	20	4320
Dermocybe cinnamomeolutea		1	1	1				1		3		1	1.3	216
Inocybe giacomi								12				12	4	564
Hebeloma spec.					1					1			0.3	20
Cortinarius spec.					1			4	5	1		9	3.3	200
Cortinarius chrysomallus								3				3	1.3	120
Inocybe cf *oreina*								1				1	0.3	46
Inocybe rhacodes								2				2	0.6	110
Russula norvegica								4				4	1.3	1776

spreading over the year of visits and on periodicity and persistence of the species involved.

f) The temporal frequency of sporocarps per month (TFCm) is also not a good importance value. However, it is a better expression of the chance to find a species throughout the year since it is less dependent on duration of research. It was used as additional character by Arnolds (1981).

g) The temporal frequency of sporocarps per year (TFCy) is a simple quantitative measure for the rate of annual fluctuation of a species. It has been advocated as additional parameter by Barkman (1976) and Arnolds (1981).

An example of a survey of mycocoenological relevés in a single plot is presented in Table V. The author (Senn-Irlet 1987) used as synthetical measures the sporocarp densities in the three years of research, the average density per year and the total dry weight in three years. The maximum density of sporocarps per 1000 m^2 would amount to e.g. 1260 for *Rickenella mellea* (Sing. & Clémençon) Lamoure and 270 for *Laccaria laccata* var. *tatrensis* Sing.

2.9.2. *Synthesis of mycocoenological data for comparison between sets of plots*

The methods for synthesis of mycocoenological data for the comparison of vegetation types or the distinction of mycocoena are similar to the procedures in the Braun-Blanquet approach in phytocoenology (for a comprehensive survey, see e.g. Westhoff and Maarel 1973, Mueller-Dombois and Ellenberg 1974). The most important synthetic character is the presence degree (or constancy when plots are of equal size), viz. the percentage of plots of a coenon in which a species is present. It is the primary criterion to determine the degree of fidelity, being an expression of the diagnostic value (character, differential, accompanying and accidental species) of a species (Braun-Blanquet, 1964). It has been used as an important criterion in most mycocoenological studies where several vegetation types were compared, e.g. by Wilkins *et al.* (1937, 1939), Darimont (1973), Lisiewska (1974), Winterhoff (1975), Arnolds (1981), Jansen (1984).

The second criterion is some averaged synthetic measure of abundance (see § 2.9.1), for instance the average maximum density of sporocarps per visit (aDCv) (e.g. Barkman 1976a, Arnolds 1981, Jansen, 1984). Often some indication of the temporal frequency is added, for instance the average annual frequency (aTFCy) (Arnolds 1981). The sum of total temporal frequency (TFC) was used by several authors from the Polish school

(Rudnicka-Jezierska 1969, Bujakiewicz 1969, Lisiewska 1974). It has the same draw-backs as importance value as the total temporal frequency in each plot (see § 2.9.1).

2.9.3. *Numerical analysis of mycocoenological data*

From the preceding paragraphs it appears that the analysis of mycocoenoses offers specific problems, which have to be solved by a methodological approach somewhat different from phytocoenology. However, the principles of data processing are fundamentally similar in these fields of vegetation science, for instance regarding techniques for automatic tabulation of data, ordination techniques, cluster analyses, etcetera. Therefore, the reader is referred to handbooks on this subject, e.g. Whittaker (1973) and Jongman *et al.* (1987). It is striking that automatic data processing and complex numerical methods have been used rarely in mycocoenology. This is caused, among other things, by the relatively small sets of plots, usually treated in mycocoenological studies. Examples of mycocoenological studies with application of modern numerical techniques are those by Senn-Irlet (1987) and Villeneuve *et al.* (1989).

2.10. Classification of macrofungocoenoses

2.10.1. *Different approaches*

Since the status of fungus communities is strongly disputed (§ 2.4) it is not surprising that the opinions on classification of macrofungocoenoses are widely different as well. Three main approaches exist: (a) the biocoenological or holistic approach, considering macrofungocoenoses as artificial units, constituting parts of biocoenoses, and regarding a formal classification undesirable in most cases; (b) the mycosyntaxonomical approach, aimed at the independent classification of macrofungocoenoses, often heterogeneous in way of habitat exploitation; (c) the mycosynusial approach, aimed at the independent classification of macrofungosocieties.

2.10.2. *Biocoenological (holistic) approach*

The adherents of this approach, including most authors of the Polish and Dutch school, prefer the incorporation of mycocoenological results in the existing classification of biocoenoses (in practice at present phytocoenoses). Besides an informal system of substrate-niche groups (see 2.4.4) may be created. Some independent mycocoenoses from habitats without plant cover may be distinguished at a syntaxonomic level, e.g. of recently burnt sites. They reject a formal hierarchical classification of both mycocoenoses and mycosocieties next to a classification of phytocoena because:

a) Macrofungi are heterogeneous in ecological and taxonomic respect (neither taxonomic nor functional merocoenoses, cf. Barkman 1973: 444) and represent only part of the heterotrophic compartment of biocoenoses.
b) The selection of plots is almost universally based on characteristics of phytocoenoses or environments, not on mycocoenoses themselves (see § 2.5).
c) Many macrofungi are ecologically and geographically dependent on certain green plants. The classification of symbiotic organisms into different systems is difficult to defend (Arnolds 1988a).
d) It is only possible to propose a formal system of coena after thorough comparative study of many macrofungocoenoses in large areas. At the moment the knowledge is only fragmentary.
e) Results so far show high correlations between classifications of phytocoenoses and macrofungocoenoses (Arnolds 1981: 249), sometimes even on the level of subassociations (Pirk 1948). Therefore hardly any information is gained by the creation of a mycocoenological system.
f) A formal system of mycocoena is a nomenclatural burden, only accessible to a few specialists.

Instead, macrofungi can be included in the lists of diagnostic taxa of phytocoena (e.g. Westhoff and Held 1969, Bon and Géhu 1973) and data on species diversity, sporocarp productivity and importance of functional groups can be included in descriptions of texture, structure and synecology of plant communities (e.g. Arnolds 1981). Descriptive surveys of characteristic macrofungi in various plant communities in larger areas (without quantitative data) were produced by Bon and Géhu (1973) for France, Kreisel and Dörfelt (1985) for the German Democratic Republic and Arnolds (1988c) for the Netherlands. For the ecological characterization of macrofungocoenoses the concepts of ecological groups or niche-substrate groups (Arnolds 1981, 1987, Barkman 1987, as 'guilds') may be used (see § 2.4.4).

2.10.3. *Mycosyntaxonomic approach*

Although most mycocoenologists analyse complete mycocoenoses, including all functional groups, formal classification of mycocoenoses is exceptional, apparently because of their evident ecological heterogeneity. However, some authors made formal descriptions of the terrestrial compartments, including both mycorrhizal and saprophytic fungi, leaving only the lignicolous fungi out of consideration. Examples are Darimont (1973) in his hierarchical classification of forest mycocoenoses in Belgium, Šmarda (1972) in Czechoslovakia and Bon (1981) in France.

2.10.4. *Mycosynusial approach*

The mycosynusial approach has been more successful than the syntaxonomic approach. The first mycosynusial 'association' (in fact union, see Wilmanns 1970) was described by Pirk and Tüxen in 1949 from excrements as Coprinetum ephemeroidis. At first synusiae were mainly described from special microhabitats, such as trunks and stumps of trees and burnt sites. In recent years also a number of ecotmycorrhizal synusiae were described, to begin with Šmarda (1972, including terrestrial saprophytes but always named after mycorrhizal species), e.g. the union Boleto aerei-Russuletum luteotactae. Synusiae of terrestrial saprophytes have been rarely described thus far. Surveys of macrofungosynusiae are given by Kreisel and Dörfelt (1985) and Fellner (1988); a detailed synopsis of synusiae on wood by Runge (1980). Part of the latter system is presented in chapter 6.2.

Macrofungosynusiae are distinguished either on the basis of analysis of many microplots in special microhabitats, selected on the basis of environment and substrate (e.g. a number of stumps of *Fagus*), or on the basis of the results of large mycocoenological plots, where a — largely hypothetical — distinction is made between different functional groups, e.g. mycorrhizal fungi versus terrestrial saprophytes (see § 2.2.2). The fundamental and practical disadvantages of such a system were mentioned before (§ 2.4.3).

A less formal and more flexible alternative is the distinction of niche-substrate groups, described in § 2.4.4. An example of niche-substrate groups in grasslands is given in chapter 5.

References

Andersson, O. 1950. Larger fungi on sandy grass heaths and sand dunes in Scandinavia. Bot. Notiser, suppl. 2: 1—89.

Apinis, A. E. 1958. Distribution of microfungi in soil profiles of certain alluvial grasslands. p. 83—90. *In*: R. Tüxen (ed.) Angewandte Pflanzensoziologie. Ber. Intern. Symposium Pflanzensoziologie — Bodenkunde 1956, Stolzenau/Weser. J. Cramer, Weinheim.

Arnolds, E. J. M. 1974. De oecologie en geografische verspreiding van *Hygrophorus* subg.

Hygrotrama, *Cuphophyllus* and *Hygrocybe* in Nederland. Report Inst. Syst. Plantkunde, University of Utrecht. 13 Tables (photo-offset).

Arnolds, E. 1980. De oecologie en sociologie van Wasplaten. Natura 77: 17—44.

Arnolds, E. 1981. Ecology and coenology of macrofungi in grasslands and moist heathlands in Drenthe, the Netherlands. Vol. 1 Biblthca mycol. 83. J. Cramer, Vaduz.

Arnolds, E. 1982. Ecology and coenology of macrofungi in grasslands and moist heathlands in Drenthe, the Netherlands. Part 2. Autecology — Part 3. Taxonomy. Biblthca mycol. 90. J. Cramer, Vaduz.

Arnolds, E. 1984. Standaardlijst van Nederlandse macrofungi. Coolia 26, suppl. Nederlandse Mycologische Vereniging, Baarn.

Arnolds, E. 1985. Typen van veranderingen in de paddestoelenflora en methoden om deze te onderzoeken. Wetensch. Meded. K.N.N.V. 167: 6—11.

Arnolds, E. 1988a. Status and classification of fungal communities. p. 153—165. *In*: J. J. Barkman, and K. V. Sykora (eds.) Dependent Plant Communities. Academic Publishing, The Hague.

Arnolds, E. 1988b. Dynamics of macrofungi in two moist heathlands in Drenthe, the Netherlands. Acta bot. Neerl. 37: 291—305.

Arnolds, E. 1988c. The Netherlands as an environmental for agarics and boleti. p. 6—29. *In*: C. Bas, Th. W. Kuyper, M. E. Noordeloos, and E. C. Vellinga (eds.) Flora Agaricina Neerlandica 1. A. A. Balkema, Rotterdam, Brookfield.

Bach, E. 1956. The Agaric *Phaeolepiota aurea*: Physiology and Ecology. Thesis Copenhagen.

Barkman, J. J. 1964. Paddestoelen in Jeneverbesstruwelen. Coolia 11: 4—29.

Barkman, J. J. 1970. Enige nieuwe aspecten inzake het probleem van synusiae en microgezelschappen. Mis. Papers Landbouwhogeschool Wageningen 5: 85—116.

Barkman, J. J. 1973. Synusial approaches to classification. p. 437—491. *In*: R. Whittaker (ed.) Handbook of vegetation science 5. Ordination and classification of vegetation. Junk, The Hague.

Barkman, J. J. 1976a. Algemene inleiding tot de oecologie en sociologie van macrofungi. Coolia 19: 57—66.

Barkman, J. J. 1976b. Terrestrische fungi in jeneverbesstruwelen. Coolia 19: 94—110.

Barkman, J. J. 1984. Biologische minimumarealen en de eilandtheorie. Vakbl. Biol. 64: 162—167.

Barkman, J. J. 1987. Methods and results of mycocoenological research in the Netherlands. p. 7—38. *In*: G. Pacioni (ed.) Studies on fungal communities. University of l'Aquila, Italy.

Benjamin, R. K. 1973. Laboulbeniomycetes. p. 223—246. *In*: G. C. Ainsworth, F. K. Sparrow, and A. S. Sussman (eds.) The Fungi: An Advanced Treatise IV A. Academic Press, London, New York.

Bohus, G., and M. Babos. 1960. Coenology of terricolous macroscopic fungi of deciduous forests. Bot. Jahrb. 80: 1—100.

Bon, M. 1981. Lactarietum lacunarum, nouvelle association fongique des lieux inondables. Docum. mycol. 11: 19—28.

Bon, M., and J. M. Géhu. 1973. Unités superieures de végetation et récoltes mycologiques. Docum. mycol. 2: 1—40.

Boois, H. M. de. 1976. Fungal development on oak leaf litter and decomposition potentialities of some fungal species. Rev. Ecol. Biol. Sol. 13: 437—448.

Bandsberg, J. W. 1966. A study of fungi associated with the decomposition of coniferous litter. Thesis Washington State University, Washington D.C.

Braun-Blanquet, J. 1964. Pflanzensoziologie (3. Aufl.) Springer, Wien, New York.

Bujakiewicz, A. 1969. Udzial grzybow wyzszych w lasach legowych i olesach puszczy bukowej pod Szczecinem. Badania Fizjograficzne nad Polska zachodnia 23, Serie B-Biologica: 61—96.

Carbiener, R., N. Ourisson, and A. Bernard. 1975. Erfahrungen über die Beziehungen zwischen Grosspilzen und Pflanzengesellschaften in der Rheinebene und den Vogesen. Beitr. naturk. Forsch. Südw. -Dtl. 34: 37—56.

Cooke, R. 1977. The biology of symbiotic fungi. John Wiley and Sons, Chichester, New York, Brisbane, Toronto.

Cooke, R. C., and A. D. M. Rayner. 1984. Ecology of saprotrophic fungi. Langman, London, New York.

Cooke, W. Bridge. 1948. A survey of literature on fungus sociology and ecology. Ecology 29: 376—382.

Cooke, W. Bridge. 1955. Fungi, Lichens and Mosses in relation to Vascular Plant Communities in Eastern Washington and adjacent Idaho. Ecological Monographs 25: 119—180.

Corner, E. J. H. 1968. A monograph of *Thelephora* (Basidiomycetes). Beih. Nova Hedw. 27. J. Cramer, Lehre.

Darimont, F. 1973. Recherches mycologiques dans les forêts de Haute Belgique. Mém. Inst. Roy. Sc. nat. Belg. 170.

Dickinson, C. H., and G. J. F. Pugh (eds.). 1974. Biology of Plant Litter Decomposition. Vols. 1 and 2. Academic Press, London, New York.

Eckblad, F. -E. 1978. Sopp Økologi. Universitetsforlaget, Oslo.

Einhellinger, A. 1969. Die Pilze der Garchinger Heide. Ein Beitrag zur Mykosoziologie der Trockenrasen. Ber. Bayer. bot. Ges. 41: 79—130.

Einhellinger, A. 1973. Die Pilze der Pflanzengesellschaften des Auwaldgebiets der Isar zwischen München und Grüneck. Ber. Bayer. bot. Ges. 44: 5—100.

Einhellinger, A. 1976. Die Pilze in primären und sekundären Pflanzengesellschaften oberbayerischer Moore. Teil I. Ber. Bayer. bot. Ges. 47: 75—149.

Einhellinger, A. 1977. Die Pilze in primären und sekundären Pflanzengesellschaften oberbayerischer Moore. Teil II. Ber. Bayer. bot. Ges. 48: 61—146.

Favre, J. 1948. Les associations fongiques des hauts-marais jurassiens et de quelques régions voisines. Matér. Flore cryptogam. Suisse 10 (3), Bern.

Favre, J. 1955. Les champignons supérieurs de la zone alpine du Parc National Suisse. Ergebn. wiss. Unters. Schweiz. Nat. Parks 5 (N.F.) 33: 1—212.

Favre, J. 1960. Catalogue descriptif des champignons supérieurs de la zone subalpine du Parc National Suisse. Ergebn. wiss. Unters. Schweiz. Nat. Parks 6 (N.F.) 42: 321—610.

Fellner, R. 1987. Principles of the arrangement of syntaxonomic classification of mycocoenoses. p. 231—245. *In*: G. Pacioni, (ed.) Studies on fungal communities, University of l'Aquila, Italy.

Fellner, R. 1988. Poznámsky k mykocoenologické syntaxonomii 2. Rrehld syntaxonomické klasifikace mykocenoz respektující zásadu jednoty substráta a trofismu (Notes to mycocoenological syntaxonomy. 2. The survey of the syntaxonomic classification of mycocoenoses taking into account the principle of the unity of the substratum and trophism). Ceská Mykologie 42: 41—51.

Frankland, J. C., J. N. Hedger, and M. J. Swift (eds.). 1982. Decomposer basidiomycetes: their biology and ecology. Cambridge University Press, Cambridge.

Guminska, B. 1962. Mikoflora lasów bukowych Rabsztynai Maciejowej (Studium florystyczno-ekologiczne). Monogr. bot. 13: 3—85.

Guminska, B. 1976. Macromycetes of meadows in Pieniny National Park. Acta mycol. 12: 3—75.

Haas, H. 1932. Die bodenbewohnenden Grosspilze in den Waldformationen einiger Gebiete von Württemberg. Beih. bot. Centralbl. 50 B: 35—134.

Haas, H. 1972. Beiträge zur Kenntnis der Pilzflora im Raum zwischen Brigach, Esbach und Prim. Schr. Vereins Geschichte und Naturgeschichte der Baar 29: 145—201.

Harley, J. L., and S. E. Smith. 1983. Mycorrhizal symbiosis. Academic Press, London, New York.

Heinemann, P. 1956. Les landes à *Calluna* du district picardo-brabançon de Belgique. Vegetatio 7: 99—147.

Hendriks, J. M. C. 1976. Kartering van paddestoelen in het Mantinger Zand. Coolia 19: 111—117.

Höfler, K. 1937. Pilzsoziologie. Ber. Deutsch. bot. Ges. 55: 606—622.

Höfler, K. 1954. Ueber Pilzaspekte. Vegetatio 5—6: 373—380.

Hueck, H. J. 1953. Myco-sociological methods of investigation. Vegetatio 4: 84—101.

Ingold, C. F. 1975. An illustrated guide to aquatic and water-borne Hyphomycetes (fungi imperfecti) with notes on their biology. Freshwater Biological Association, Ambleside, Scientific Publ. No. 30.

Jahn, H. 1979. Pilze die an Holz wachsen. Baranek and Frost, Herford.

Jansen, A. E. 1984. Vegetation and macrofungi of acid oak woods in the north-east of the Netherlands. Agriculture Research Reports 923. Pudoc, Wageningen.

Jongman, R. H. G., C. J. F. ter Braak, and O. F. R. van Tongeren. 1987. Data analysis in community and landscape ecology. Pudoc, Wageningen.

Kalamees, K. 1965. On problems and methods in mycosociology. Akad. Nauk. Eston. S.S.R. Tartu: 14—22.

Kalamees, K. 1968. Mycocoenological methods based on investigations in the Estonian forests. Acta mycol. 4: 327—334.

Kalamees, K. 1979. The role of fungal groupings in the structure of ecosystems. Eesti NSV Teaduste Akadeemia Toimetised 28 (3) Bioloogia: 206—213.

Kreisel, H. 1957. Die Pilzflora des Darss und ihre Stellung in der Gesamtvegetation. Beih. Rep. Spec. Nov. Veg. 137. Beitr. Vegetationskunde 2: 110—183.

Kreisel, H., and H. Dörfelt. 1985. Pilzsoziologie. p. 67—95. *In*: Michael, Hennig, and Kreisel (eds.) Handb. Pilzfr. 4 (3. Aufl.) Gustav Fischer, Jena.

Kreisel, H., and G. Ritter. 1985. Ökologie der Grosspilze. p. 9—47 *In*: Michael, Hennig, and Kreisel (eds.) Handb. Pilzfr. 4 (3. Aufl.) Gustav Fischer, Jena.

Lange, M. 1948. The agarics of Maglemose. Dansk bot. Ark. 13: 1—141.

Leischner-Siska, E. 1939. Zur Soziologie und Ökologie der höheren Pilze. Untersuchung der Pilzvegetation in der Umgebung von Salzburg während des Maximalaspektes 1937. Beih. bot. Centralblatt 59: 359—429.

Lisiewska, M. 1974. Macromycetes of beech forests within the eastern part of the *Fagus* area in Europe. Acta mycol. 10: 3—72.

Lisiewska, M. 1978. Macromycetes na the zespolów lesnych Swietokrzyskiego Parku Narodowego. Acta mycol. 14: 163—191.

Marks, G. C., and T. T. Kozlowski. 1973. Ectomycorrhizae, their ecology and physiology. Academic Press, New York, London.

Moser, M. 1949. Untersuchungen über den Einfluss von Waldbränden auf die Pilzvegetation I. Sydowia 3: 336—383.

Mueller-Dombois, D., and H. Ellenberg. 1974. Aims and Methods of Vegetation Ecology. J. Wiley and Sons, New York, London, Sydney, Toronto.

Murakami, Y. 1987. Spatial distribution of *Russula* species in *Castanopsis cuspidata* forest. Trans. Br. mycol. Soc. 89: 187—193.

Neergaard, P. 1977. Seed Pathology, Vols. 1, 2. Macmillan, London, Basingstoke.

Nespiak, A. 1971. Grzyby wyzsze regla görnego w Karkonoszach. (Die Pilze in den Piceetum hercynicum in Karkonosze). Acta mycol. 7: 87—98.

Ohenoja, E. 1978. Mushrooms and mushroom yields in fertilized forests. Ann. bot. Fennici 15: 38—46.

Ohenoja, E. 1973. Documentation on the research into the fruiting body production of forest fungi. Memoranda Soc. Fauna Flora Fennica 59: 113—115.

Parker-Rhodes, A. F. 1951. The Basidiomycetes of Skokholm Island VII. Some floristic and ecological calculations. New Phytologist 50: 227—243.

Parker-Rhodes, A. F. 1955. The Basidiomycetes of Skokholm Island XIII. Correlation with the chief plant associations. New Phytologist 54: 259—276.

Parker-Rhodes, A. F. 1955. The Basidiomycetes of Skokholm Island XIII. Correlation with the chief plant associations. New Phytologist 54: 259—276.
Petersen, P. R. M. 1970. Danish fireplace fungi. An ecological investigation on fungi on burns. Dansk bot. Ark. 27 (3): 1—97.
Pirk, W. 1948. Zur Soziologie der Pilze im Querceto-Carpinetum. Z. Pilzk., N.F. 1: 11—20.
Pirk, W., and R. Tüxen. 1949. Das Coprinetum ephemeroidis, eine Pilzgesellschaft auf frischem Mist der Weiden im mittleren Wesertal. Mitt. flor.-soz. Arbeitsgem., N.F. 1: 1—7.
Pirk, W., and R. Tüxen. 1957. Höhere Pilze in nw. -deutschen Calluna-Heiden (Calluneto-Genistetum typicum). Mitt. flor.-soz. Arbeitsgem., N.F. 6/7: 127—129.
Rammeloo, J., and L. Vanhecke. 1979. *Hirneola auricula-judae* (Judasoor) in België: proeve tot verklaring van het verspreidingspatroon. Dumortiera 13: 10—25.
Reid, D. A. 1974. Changes in the British Macromycete Flora. p. 79—85. *In*: D. L. Hawksworth (ed.) The changing flora and fauna of Britain. Academic Press, London, New York.
Reijnders, A. F. M. 1968. Terugblik op 1967. Coolia 13: 111—114.
Richardson, M. J. 1970. Studies on *Russula emetica* and other agarics in a Scots pine plantation. Trans. Br. mycol. Soc. 55: 217—229.
Rudnicka-Jezierska, W. 1969. Higher fungi of the inland dunes of the Kampinos Forest near Warsaw. Monogr. bot. 30: 3—116.
Runge, A. 1980. Pilz-Assoziationen auf Holz in Mitteleuropa. Z. Mykol. 46: 95—102.
Sadowska, B. 1973. Preliminary evaluation of the productivity of fungi (Agaricales and Gasteromycetes) on the Kazun meadows. Acta mycol. 9: 91—100.
Sanders, F. E., B. Mosse, and P. B. Tinker (eds.). 1975. Endomycorrhizas. Academic Press, London, New York, San Francisco.
Schenck, N. C. (ed.). 1982. Methods and principles of mycorrhizal research. American Phytopathological Society, St. Paul.
Senn-Irlet, B. J. 1987. Ökologie, Soziologie und Taxonomie alpiner Makromyzeten (Agaricales, Basidiomycetes) der Schweizer Zentralalpen. Diss. Bern.
Sewell, G. W. F. 1959a. Studies of fungi in a *Calluna*-heathland soil, I. — Vertical distribution in soil and on root-surfaces. Trans. Br. mycol. Soc. 42: 343—353.
Sewell, G. W. F. 1959b. The ecology of fungi in *Calluna*-heathland soils. New Phytologist 58: 5—15.
Shantz, H. L. and R. L. Piemeisel. 1917. Fungus fairy rings in eastern Colorado and their effect on vegetation. J. agric. Research 11: 191—145.
Šmarda, F. 1968. Kriterien der soziologischen Bewertung der Pilze. Česká Mykol. 22: 114—120.
Šmarda, F. 1972. Die Pilzgesellschaften einiger Laubwälder Mährens. Prirodoved. Pr. Ustavu CSAV Brne 6: 1—53.
Sparrow, F. K. 1960. Aquatic Phycomycetes (2nd ed.). The University of Michigan Press, Ann Arbor.
Swift, M. J., O. W. Heal, and J. M. Andersson. 1979. Decomposition in terrestrial ecosystems. Studies in Ecology 5. Blackwell, Oxford, London, Edinburgh, Melbourne.
Thoen, D. 1970—1971. Etude mycosociologique de quelques associations forestières des districts picardo-brabančon, mosan et ardennais de Belgique. Bull. Rech. agronom. Gembloux 5: 309—326; 6: 215—243.
Thoen, D. 1976. Een onderzoek naar de mycocoenosen van de sparrenaanplantingen in de Zuidelijke Ardennen. Coolia 19: 67—71.
Thoen, D. 1977. Contribution à l'étude des mycocénoses des pessières artificielles d'Ardenne méridionale. Diss. Université Arlon.
Torkelsen, A. E. 1968. The genus *Tremella* in Norway. Nytt Mag. Bot. 15: 225—239.
Villeneuve, N., M. M. Grandtner, and J. A. Fortin. 1889. Frequency and diversity of ectomycorrhizal and saprophytic macrofungi in the Laurentide Mountains of Québec. Can. J. Bot. 67: 2616—2629.

Vries, B. de. 1990. On the quantitative analysis of wood-decomposing macrofungi in forests. I. Wageningen Agricultural Papers 90 (b): 93—101.
Warcup, J. H. 1951. The ecology of soil fungi. Trans. Br. mycol Soc. 34: 376—399.
Warcup, J. H., and P. H. B. Talbot. 1962. Ecology and identity of mycelia isolated from soil. Trans. Br. mycol. Soc. 45: 495—518.
Westhoff, V., and E. van der Maarel. 1973. The Braun-Blanquet approach. p. 617—726. *In*: R. H. Whittaker (ed.) Handbook of Vegetation Science 5. Junk, The Hague.
Whittaker, R. H. (ed.). 1973. Ordination and classification of communities. Handbook of Vegetation Science 5. Junk, The Hague.
Wilkins, W. H., E. M. Ellis, and J. L. Harley. 1937. The ecology of the larger fungi. I. Constancy and frequency of fungal species in relation to certain vegetation communities, particularly oak and beech. Ann. appl. Biol. 24: 703—732.
Wilkins, W. H., J. L. Harley, and G. C. Kent. 1939. The ecology of larger fungi. II. Ann. appl. Biol. 25: 472—489.
Wilkins, W. H., and S. H. M. Patrick. 1939. The ecology of larger fungi. III. Constancy and frequency of grassland species with special reference to soil types. Ann. appl. Biol. 26: 25—46.
Wilkins, W. H., and S. H. M. Patrick. 1940. The ecology of the larger fungi. IV. The seasonal frequency of grasslands fungi with special reference to the influence of environmental factors. Ann. appl. Biol. 27: 17—34.
Wilmanns, O. 1970. Kryptogamen-Gesellschaften oder Kryptogamen-Synusien? p. 1—6. *In*: R. Tüxen (ed.) Gesellschaftsmorphologie. Ber. Intern. Symp. Vegetationsk. Rinteln 1966.
Winterhoff, W. 1975. Die Pilzvegetation der Dünenrasen bei Sandhausen (nördl. Oberrheinebene). Beitr. naturk. Forsch. Südw. -Dtl. 34: 445—462.
Winterhoff, W. 1977. Die Pilzflora des Naturschutzgebietes Sandhausener Dünen bei Heidelberg. Veröff. Naturschutz Landschaftspflege Bad. Württ. 44/45: 51—118.
Winterhoff, W. 1984. Analyse der Pilze in Pflanzengesellschaften, insbesondere der Makromyzeten. p. 227—248. *In*: R. Knapp (ed.) Sampling methods and taxon analysis in vegetation science. Dr. W. Junk Publishers, The Hague, Boston, Lancaster.
Witkamp, M. 1960. Seasonal fluctuations of the fungus flora in mull and mor of an oak forest. Itbon Meded. 46: 1—51.
IJpelaar, P. 1985. Veranderingen in de mycoflora van eikenbossen op arme zandgrond. p. 70—73. *In*: E. Arnolds (ed.) Veranderingen in de paddestoelenflora (mycoflora). Wetensch. Meded. K.N.N.V. 167, Hoogwoud.

3. Macrofungi on soil in deciduous forests

ANNA BUJAKIEWICZ

Abstract

A survey is given on the macrofungocoenoses of the following Central European deciduous forest community types: Alnion glutinosae, Alno-Padion, Tilio-Acerion, Carpinion betulae, Fagion sylvaticae, Quercion petraeo-pubescentis, and Quercion robori-petraeae. Characteristic differences of the species composition and of the contribution of litter saprotrophs, humicolous saprotrophs and mycorrhizal fungi to the fungocoenoses are discussed. The ecological and sociological indicative value of macrofungi is demonstrated. Seasonal aspects and mosaic patterns are mentioned. Some comparisons are made with mycocoenoses of coniferous forests and with mycocoenoses of forests in Northern Europe and in the eastern part of North America.

3.1. Introduction

Forests are the most appropriate of all natural habitats for studying connections between macrofungi and plant community. However, an intricate structure of forests brings about extreme difficulties in the definition and designation of various fungal groupings occurring within the forest biocoenosis.

The role of macrofungi growing on soil[1] is highly important and significant in the formation of the broad-leaved, deciduous forests. They provide much better conditions for an active breakdown of litter than pure coniferous stands.

Litter saprotrophs in a broad-leaved forest are represented by rich populations of lignin and cellulose decomposing fleshy fungi, such as

[1] Terrestrial fungi considered in this chapter include biotrophic (mycorrhizal) fungi and saprophytic fungi, growing on humus and litter.

W. Winterhoff (ed.), Fungi in Vegetation Science, 49—78.

Clitocybe, Lepista, Mycena, Psathyrella and many others that overgrow and bind fallen leaves.

Humicolous saprotrophs decompose older leaves and unrecognizable plant debris in the fermentation and humus layers under the litter. This group includes a fair number of fungi such as *Agaricus, Calocybe, Conocybe, Entoloma, Lepiota, Macrolepiota, Melanophyllum, Mitrophora, Rhodocybe* and *Volvariella.* Delimitation of these two groups of saprophytes is very vague. They both play a significant role in the complete breakdown of leafy litter and lead to formation of a mull type of humus.

Mycorrhizal fungi are linked with certain host trees and are responsible for biomas production. They probably participate very little in decomposition of organic matter in the soil but their mycelium penetrates very deep into mineral soil. Many fungi, considered for a long time as saprophytic, turned out to be mycorrhizal (*Laccaria*, many species of *Hebeloma, Inocybe*). An intensive research on mycorrhiza observed at present will bring a new light to this problem (Agerer 1985, 1987). Trappe (1962) listed taxa of mycorrhiza formers for the majority of forest tree species. The main genera of mycorrhizal broad-leaved forest fungi are: *Amanita, Boletus, Cortinarius, Lactarius, Russula* and *Tricholoma.*

Terrestrial forest fungi, however widely distributed, have specific habitat preferences for different types of litter, for soil and air moisture, for types of soil parent material, soil acidity etcetera. They are sensitive ecological indicators for certain factors in a forest biocoenosis and react strongly to environmental changes.

Many authors have discussed the correlation between macrofungi and deciduous forest associations (Bohus and Babos 1960, 1967, Darimont 1973, Gumińska 1962, Haas 1932, Jahn *et al.* 1967, Lisiewska 1965, 1972, 1974, Ławrynowicz 1973, Nespiak 1959, 1962, Sałata 1972, Wojewoda 1975) and some have considered the problem of the response of fungi to the pH of the soil (Bohus 1984, Tyler 1985). All these papers prove that this correlation does exist and that the agreement with terrestrial mycoflora in closely related forest types is striking (Lisiewska 1972, 1974) despite geographical distances (Bujakiewicz 1977, 1985).

A theoretical discussion on terrestrial fungi and the problem of synusial grouping (mycocoenoses) was recently given by Arnolds (1981), Barkman (1973), Kalamees (1971, 1979) and Winterhoff (1984). Noteworthy is an attempt to enumerate characteristic fungi on the level of higher syntaxa of the main plant communities in France (Bon and Géhu 1973). There has also been an approach to classify the groupings of terrestrial forest fungi in East Belgium by Darimont (1973) and a similar attempt introduced by Šmarda (1972) in the Moravian forests. Both authors define the groupings by their fungal components and dominants.

This is an attempt to show the ecological amplitude of chosen forest macrofungi mainly in Central Europe (Table I and II).

More precise data, including the author's own investigations for the Alnion glutinosae and the Alno-Padion, allow to consider this problem against the range of forest associations (Table I). Part A of this table shows fungi species exclusive for the forest associations or connecting two or more forest associations, part B shows fungi with broader ecological range and part C shows the scale of saprophytic fungi growing on fallen twigs, leaves and fruits.

Table II gives an overall picture of the diversity of forest macrofungi and their habitat preferences. The material was taken from different areas and forests on various soils and were put together in one syntaxon (alliance).[2] The question of an attachment of macrofungi to forest types is discussed at the level of syntaxa higher in rank, i.e. alliances, orders and classes. All syntaxa, in both tables, are arranged according to humidity and fertility of habitat sequence. The first part of the Table II (part A) shows the range of macrofungi against the whole scale of forests analised, part B shows the attachment of fungi to certain alliances that occur in more than one syntaxon but are always absent in certain others and part C shows the scale of fungi with a broad ecological spectrum but inclined to a certain syntaxa.

3.2. Alnion glutinosae (Malc. 1929) Meijer-Dress 1936

The most striking feature of the Carici elongatae-Alnetum s.l., the main forest type of that syntaxon, is the mosaic structure of habitats brought about by fluctuations of the ground water level.

On hummocks, 1 to 2 m high, formed by adventitious roots of *Alnus glutinosa*, covered with acidic humus and with mosses, many bryophilous fungi occur (*Rickenella fibula, R. setipes*), various species of *Galerina*, litter saprophytic fungus Mycena sanguinolenta and some mycorrhiza formers with *Alnus*, such as *Lactarius obscuratus* that grows on bare humus on hummock escarpments or among mosses.

On the organic peaty soil of the boggy hollows a fairly large number of fungi occurs, mainly mycorrhizal with alder. Some follow alder roots and spread over the boggy ground (*Naucoria escharoides, N. scolecina*), others grow on plant debris in an open bog inbetween hummocks (*Cortinarius alneus, Lactarius omphaliformis*). They form well-recognized fungus aspects in late summer and autumn (Bujakiewicz 1973). Typical acidophilic fungi are

[2] For comparison only papers were taken that considered well-defined forest associations and presented mycocoenological tables with appropriate data.

Table I. Ecological amplitude of macrofungi in the forests of the Alnion glutinosae and the Alno-Padion

Alliance		ALNION		ALNO-PADION							
Association		Spc	CeA	FA	FUC		CrF	AF	AP	CA	VU
					ch	t					
Number of localities		1	8	7	5	6	2	1	5	1	1
Number of plots		2	12	22	16	19	5	3	26	5	2
Duration of studies (in years)		2	2—5	2—5	3	1—3	3	3	1—2	4	3
L.f. Part A											
M	*Inocybe lanuginosa*	1a									
M	*Russula versicolor*	1a									
M	*Naucoria langei*	1n	IIIn	Ir							
M	*Hebeloma pusillum*	1n		Ir							
S	*Conocybe striaepes*		IIr								
S	*Conocybe sordida*		Ir								
M	*Cortinarius alneus*		Ir								
M	*Naucoria alnetorum*		IIIn	IIIn							
M	*Lactarius omphaliformis*		IIIn	In							
S	*Entoloma minutum*		IIr	IIr	1n						
S	*Psathyrella vernalis*		IIr	In	1a						
M	*Lactarius pyrogalus*		Ir		1n						
S	*Conocybe brunnea*			Ir							
S	*C. kuehneriana*			Ir							
S	*Entoloma icterinum*			Ir	1r						

Table I. (continued)

	Alliance	ALNION		ALNO-PADION							
	Association	Spc	CeA	FA	FUC		CrF	AF	AP	CA	VU
					ch	t					
	Number of localities	1	8	7	5	6	2	1	5	1	1
	Number of plots	2	12	22	16	19	5	3	26	5	2
	Duration of studies (in years)	2	2—5	2—5	3	1—3	3	3	1—2	4	3
M	*Inocybe calospora*			Ir	1r						
S	*Mitrophora semilibera*			Ir	1r						
S	*Calocybe gambosa*				1n						
S	*Entoloma lanicum*				1n						
S	*Langermannia gigantea*				1n						
M	*Russula lutea*				1n						
S	*Melanophyllum eyrei*				1r						
S	*Bovista pusilliformis*				1r	Ir					
S	*Entoloma clypeatum*				1r	In					
M	*Xerocomus rubellus*					IIIr					
S	*Agrocybe praecox*					IIn					
S	*Morchella esculenta*					IIr					
S	*Cystolepiota bucknallii*					Ir					
S	*Cystolepiota hetieri*					Ir					
S	*Lepista irina*					Ir	1r				
S	*Entoloma sericellum*						1r				
M	*Tricholoma myomyces*						1r				

Table I. (continued)

Alliance		ALNION		ALNO-PADION							
Association		Spc	CeA	FA	FUC		CrF	AF	AP	CA	VU
					ch	t					
Number of localities		1	8	7	5	6	2	1	5	1	1
Number of plots		2	12	22	16	19	5	3	26	5	2
Duration of studies (in years)		2	2–5	2–5	3	1–3	3	3	1–2	4	3
M	*Cortinarius uliginosus*						1r	1n			
S	*Entoloma nidorosum*						1r	1n			
S	*Rhodocybe nitellina*								1a		
M	*Inocybe calamistrata*								1n		
M	*Naucoria luteolofibrillosa*								1n		
M	*Lactarius aspideus*								1n		
S	*Entoloma lepidissimum*								1r		
M	*Naucoria suavis*								1r		
M	*Cortinarius pulchripes*									1a	
S	*Bovista aestivalis*				1r						1r
S	*Macrolepiota procera*										1a
Part B											
M	*Lactarius lilacinus*	1r	IIr	IIr							
M	*Naucoria escharoides*	1n	Va	Va	1n						
M	*Naucoria scolecina*		Va	Va	1a						
M	*Lactarius obscuratus*		IIIn	Vn	1r						

Table I. (continued)

Alliance		ALNION		ALNO-PADION							
Association		Spc	CeA	FA	FUC		CrF	AF	AP	CA	VU
					ch	t					
Number of localities		1	8	7	5	6	2	1	5	1	1
Number of plots		2	12	22	16	19	5	3	26	5	2
Duration of studies (in years)		2	2—5	2—5	3	1—3	3	3	1—2	4	3
M	*Naucoria subconspersa*		IIIn	IIIn	1r						
M	*Cortinarius bibulus*		IIr	IIIn	1r						
M	*Russula pumila*		IIr	Ir				1r	1a		
M	*Paxillus filamentosus*		Ir	Ir				1r			
M	*Cortinarius alnetorum*		Ir	IIa	1r				1r		
M	*Naucoria bohemica*			Ir							
M	*Cortinarius helvelloides*		IIIn	IIn							
S	*Lepiota cristata*		IIn	IIn	1a	IIr					
S	*Coprinus cortinatus*		Ir	Ir	1r	IIr					
S	*Psathyrella orbitarum*		Ir	Ir	1n	IIr					
S	*Conocybe arrheni*		Ir			IIr					
M	*Hebeloma sacchariolens*		Ir	IIn	1n	Ir					
M	*Inocybe fastigiata*		Ir	IIr	1r	Ir					
S	*Macroscyphus macropus*		Ir	In		Ir					
S	*Conocybe mairei*		Ir	Ir	1r	Ir	1r				
M	*Cortinarius flexipes*		Ir	Ir			1r				
M	*Inocybe geophylla* v. *geophylla*		IIn	Ir	1n		2r	1r			
S	*Entoloma infula*		In	Ir			1r	1r			
S	*Tubaria pellucida*		IIr	Ir	1n		1r				

Table I. (continued)

	Alliance	ALNION		ALNO-PADION							
	Association	Spc	CeA	FA	FUC		CrF	AF	AP	CA	VU
					ch	t					
	Number of localities	1	8	7	5	6	2	1	5	1	1
	Number of plots	2	12	22	16	19	5	3	26	5	2
	Duration of studies (in years)	2	2—5	2—5	3	1—3	3	3	1—2	4	3
S	*Melanophyllum haematospermum*				1r	IIIr	1r				
S	*Rhodophyllus radiatus*					Ir	1r				
S	*Tarzetta cupularis*			Ir	1r	IIr	1r				
S	*Entoloma strigosissimum*			Ir	1n	Ir	1r				
S	*Conocybe filaris*			Ir	1r	In					
S	*Conocybe aporos*			Ir		Ir					
S	*Entoloma juncinum*			Ir	1r				1r		
S	*Laccaria tortilis*			Ir	1r		1r		1r		
M	*Inocybe geophylla* v. *violacea*			Ir	1n		1r				1r
S	*Clitocybe fragrans*			IIr	1n		1r	1r			
S	*Lepiota clypeolaria*				1r				1r		
S	*Cystolepiota sistrata*			Ir	1r	IIr			1r		1r
S	*Phallus impudicus*			Ir	1r	Ir		1r			1r
M	*Tricholoma album*				1a				1r		1r
S	*Lepista nebularis*				1r						1r
S	*Lepista nuda*					Ir					1r
S	*Clitocybe gibba*								1r		1r
S	*Lepiota castanea*										1r

Table I. (continued)

	Alliance	ALNION		ALNO-PADION							
	Association	Spc	CeA	FA	FUC		CrF	AF	AP	CA	VU
					ch	t					
	Number of localities	1	8	7	5	6	2	1	5	1	1
	Number of plots	2	12	22	16	19	5	3	26	5	2
	Duration of studies (in years)	2	2–5	2–5	3	1–3	3	3	1–2	4	3
	Part C										
S	*Rutstroemia conformata*		Ir								
S	*Ciboria viridifusca*		Ir								
S	*Ciboria amentacea*		In	In							
S	*Rutstroemia firma*		Ir	Ir	1r	Ir					
S	*Typhula erythropus*		Ia	Ia		Ir					
S	*Peniophora erikssonii*		Ir			Ir				1r	
S	*Phaeomarasmius erinaceus*		Ir	In					1r	1n	
S	*Macrotyphula fistulosa*		In	IIa					1r	1r	
S	*Hymenoscyphus imberbis*			Ia		Ir					
S	*Mollisia melaleuca*			Ia	1a						
S	*Clitopilus hobsonii*			In							
S	*Ciboria alni*			Ir							
S	*Pezizella alniella*			Ir						1a	
S	*Simocybe centunculus*				1n	IIr					
S	*Pleurotellus chioneus*				1n		1r				
S	*Psathyrella gracilis*				1n			1r			
S	*Hymenoscyphus robergei*			IIa	1a						
S	*Psathyrella gyroflexa*			Ir	1n	Ir					

Table I. (continued)

Alliance		ALNION		ALNO-PADION							
Association		Spc	CeA	FA	FUC		CrF	AF	AP	CA	VU
					ch	t					
Number of localities		1	8	7	5	6	2	1	5	1	1
Number of plots		2	12	22	16	19	5	3	26	5	2
Duration of studies (in years)		2	2—5	2—5	3	1—3	3	3	1—2	4	3
S	*Polyporus ciliatus*			IIr		IIn	1r	1r			
S	*Simocybe rubi*			IIr	1r	IIr			1r		
S	*Pluteus phlebophorus*			IIr	1n	IIr		1r	1r		
S	*Hymenoscyphus fructigenum*			Ir	1n	IIa					1r
S	*Pluteus semibulbosus*				1n	Ir					1r
S	*Skeletocutis nivea*					In					
S	*Coprinus lagopus*		Ir	Ir	1a						
S	*Pluteus thomsonii*		Ir	Ir	1r	Ir					
S	*Psathyra squamifera*		Ir	In	1n	Ir	1n				
S	*Crepidotus pubescens*		IIr	Ir	1r	Ir		1r			
S	*Micromphale foetidum*		Ir		1n		1r	1r			
S	*Polyporus varius* v. *nummularius*		Ir			Ir	1r		1n	1r	
S	*Coprinus impatiens*		Ir	Ir	1n	Ir	1r				1r
S	*Psathyrella trepida*		Ir	Ir	1r		1r				1r
S	*Marasmius lupuletorum*		Ir	Ir	1a						1n
S	*Mycena flavoalba*		Ir	In	1n	Ir		1r			
S	*Crepidotus subsphaerosporus*		Ir	Ir	1n	IIr	2r	£R			£R
S	*Crucibulum laeve*		Ir	IIr	1n	IIr		1r	1r		
S	*Coprinus plicatilis*		Ir	IIIr	1r		1r		1r		
S	*Mycena chlorinella*		IIIr	IIIr	1r	1r	1r				
S	*Mycena stylobates*		IIIr	IIIr	1r	IIr		1r			

Table I. (continued)

	Alliance	ALNION		ALNO-PADION							
	Association	Spc	CeA	FA	FUC		CrF	AF	AP	CA	VU
					ch	t					
	Number of localities	1	8	7	5	6	2	1	5	1	1
	Number of plots	2	12	22	16	19	5	3	26	5	2
	Duration of studies (in years)	2	2–5	2–5	3	1–3	3	3	1–2	4	3
S	*Psathyrella candolleana*		IIIn	IIIa	1a	IIr	1r	1r	1r		
S	*Marasmiellus ramealis*		IVn	IIIn	1a	IIr		1r	1r		
S	*Tubaria furfuracea*		IIIr	IIIn	1n	Ir	1r		1r	1r	1r
S	*Mycena speirea*		IVn	Va	1a	IVa	2n	1n	1n	1n	1r
S	*Mycena acicula*	1r	Ir	Ir	1a	IVn	2r	1r		1r	1r
S	*Marasmius epiphyllus*	1n	IIIr	IIn	1a		2n	1r	1r		
S	*Psathyrella obtusata*	1r		Ir	1a		1r				

Explanations:
L.f. = life form: M = mycorrhizal (based mainly on Trappe 1962)
S = saprophytic (on humus and litter)
1, II . . . = constancy in phytosociological meaning
a,n,r = degree of abundance: a = abundant, n = numerous, r = rare
Alnion glutinosae: Spc = Salicetum pentandro-cinereae (Almq. 1929) Pass. 1961: Nespiak (1959); CeA = Carici elongatae-Alnetum Koch 1926: Nespiak (1959), Bujakiewicz (1969, 1973, 1987a), Friedrich (1985), Kristoffersen (unpubl.); Alno-Padion: FA = Fraxino-Alnetum Matuszk. 1952: Bujakiewicz (1967, 1973, 1987b), Friedrich (1985), Lisiewska (1978), Nespiak (1959); FUc = Ficario-Ulmetum campestris Knapp 1942 em.J.Mat. 1976, ch = chrysosplenietosum: Bujakiewicz (1973); t = typicum: Bujakiewicz (1977, 1985, unpubl.), Wojewoda (1975); CrF = Carici remotae-Fraxinetum Koch 1926: Bujakiewicz (1967) and Equiseto-Fraxinetum K.-Lund 1971: Markussen (unpubl.); AF = Alno incanae-Fraxinetum K.-Lund ap.Seib. 1969: Kristoffersen (unpubl.); AP = Alno incanae-Prunetum K.-Lund 1971: Bujakiewicz (unpubl.); CA = Caltho incanae-Alnetum (Zarz. 1963) Stuchlik 1968: Bujakiewicz (1981); VU = Violo odoratae-Ulmetum Veevers (1940) Doing 1962: Bujakiewicz (unpubl.)

Table II. Ecological amplitude of macrofungi in deciduous forests

	CLASS	ALNETEA	QUERCO-FAGETEA					QRP
	ORDER	Aln	Fagetalia				Qpp	Qrp
	ALLIANCE	Alg	A-P	T-A	Car	Fag	Qpp	Qrp
	Number of associations analized	2	8	2	2	10	2	3
	L.f. Part A							
S	*Conocybe sordida*	1						
S	*Conocybe striaepes*	1						
M	*Cortinarius alneus*	1						
M	*Russula versicolor*	1						
M	*Naucoria langei*	2	I					
M	*Hebeloma pusillum*	1	I					
M	*Naucoria alnetorum*	1	I					
S	*Conocybe arrheni*	1	I					
M	*Russula pumila*	1	II					
M	*Cortinarius helvelloides*	1	II					
M	*Paxillus filamentosus*	1	II					
S	*Psathyrella orbitarum*	1	II					
S	*Coprinus cortinatus*	1	II					
M	*Cortinarius alnetorum*	1	III					
M	*Lactarius obscuratus*	1	IV					
M	*Cortinarius bibulus*	1	IV					
M	*Lactarius lilacinus*	2	II	1				
M	*Naucoria scolecina*	1	IV		1			
S	*Psathyrella vernalis*	1	II		1			
S	*Entoloma minutum*	1	II	2				

Table II. *(continued)*

CLASS		ALNETEA	QUERCO-FAGETEA					QRP
ORDER		Aln	Fagetalia				Qpp	Qrp
ALLIANCE		Alg	A-P	T-A	Car	Fag	Qpp	Qrp
Number of associations analized		2	8	2	2	10	2	3
M	*Lactarius omphaliformis*	1	I		1			
S	*Conocybe mairei*	1	III		1	I		
S	*Tubaria pellucida*	1	III			I		
M	*Naucoria subconspersa*	1	III			I		
S	*Macroscyphus macropus*	1	II	1	2	II		
M	*Naucoria escharoides*	2	IV		1		1	
M	*Inocybe fastigiata*	1	II		2	IV	1	
S	*Lepiota cristata*	1	II	1	2	II	1	
M	*Amanita phalloides*	1			2	III	2	
M	*Lactarius subdulcis*	1			2	III	1	1
M	*Russula delica*	1			2	IV	2	1
M	*Hebeloma longicaudum*	1			2	I	1	1
M	*Lactarius pyrogalus*	1	I	1	2	II	2	1
M	*Russula ochroleuca*	1	I	1	2	II		2
S	*Clitocybe fragrans*	1	II		2	I		2
S	*Rhodophyllus radiatus*		II		2			1
S	*Cystolepiota sistrata*		IV	1	2	I		1
M	*Tricholoma myomyces*		I		2	II		1
S	*Lepiota clypeolaria*		II		2	II	1	1
M	*Xerocomus chrysenteron*		I	2	2	IV		2
M	*Russula lutea*		I		2	III	3	1
M	*Tricholoma sulphureum*		I		2	III	2	1

Table II. (continued)

CLASS		ALNETEA	QUERCO-FAGETEA					QRP
ORDER		Aln	Fagetalia				Qpp	Qrp
ALLIANCE		Alg	A-P	T-A	Car	Fag	Qpp	Qrp
Number of associations analized		2	8	2	2	10	2	3
M	*Hygrophorus eburneus*		I		2	III	2	1
S	*Lepista nebularis*		II	1	2	II	2	1
M	*Russula cyanoxantha*		I	1	2	IV	2	2
S	*Lepista nuda*		II	1	2	I	2	3
M	*Inocybe cookei*			1	2	I	2	1
M	*Lactarius blennius*			1	2	IV		1
M	*Inocybe hirtella*			1	1		2	1
M	*Russula nigricans*				2	III	2	2
M	*Russula densifolia*				2	II	1	1
M	*Tricholoma ustale*				1	III	1	1
M	*Russula fellea*				2	III	1	1
M	*Russula vesca*				1	III	3	1
S	*Stropharia squamosa*				1	II		1
M	*Pseudocraterellus cinereus*				1	III		1
M	*Lactarius pallidus*				1	III		1
M	*Lactarius piperatus*				2	II		1
M	*Boletus edulis*				2	III		1
S	*Otidea leporina*				2	I		1
M	*Lactarius camphoratus*				2	III		2
M	*Lactarius chrysorheus*				2	I	2	1
M	*Lactarius quietus*				2	I	3	2
M	*Tylopilus felleus*				2	I		3
M	*Russula parazurea*				1		1	2
M	*Russula olivacea*					III	1	1

Table II. (continued)

CLASS		ALNETEA	QUERCO-FAGETEA					QRP
ORDER		Aln	Fagetalia				Qpp	Qrp
ALLIANCE		Alg	A-P	T-A	Car	Fag	Qpp	Qrp
Number of associations analized		2	8	2	2	10	2	3
M	*Cortinarius nemorensis*					II	1	1
P	*Cordyceps ophioglossoides*					I		1
M	*Cortinarius decipiens*					III		2
M	*Elaphomyces granulatus*					I	2	1
S	*Laccaria proxima*							3
M	*Russula emetica*							3
S	*Clitocybe diatreta*							2
S	*Psathyrella frustulenta*							2
S	*Psathyrella fulvescens*							2
S	*Tephrocybe tylicolor*							2
Part B								
S	*Entoloma strigosissimum*		III					
S	*Bovista pusilliformis*		III					
S	*Conocybe aporos*		II					
S	*Bovista aestivalis*		I					
M	*Cortinarius pulchripes*		I					
S	*Entoloma lanicum*		I					
S	*Langermannia gigantea*		I					
M	*Naucoria luteolofibrillosa*		I					
M	*Inocybe calospora*		II	1				

Table II. (continued)

CLASS		ALNETEA	QUERCO-FAGETEA					QRP
ORDER		Aln	Fagetalia				Qpp	Qrp
ALLIANCE		Alg	A-P	T-A	Car	Fag	Qpp	Qrp
Number of associations analized		2	8	2	2	10	2	3
S	*Entoloma icterinum*		II	1				
S	*Cystolepiota bucknallii*		I	1				
S	*Cystolepiota hetieri*		I	1				
S	*Lepista irina*		II	1	1			
M	*Naucoria bohemica*		II		2			
S	*Conocybe filaris*		II		1			
S	*Melanophyllum haematospermum*		II		2			
M	*Xerocomus rubellus*		I		2			
S	*Melanophyllum eyrei*		I		2			
M	*Inocybe maculata*			1	1	I		
S	*Agrocybe erebia*			2	2	I		
M	*Pseudocraterellus sinuosus*			1	1	I		
S	*Coprinus picaceus*				1	I		
S	*Pholiota lenta*				1	II		
M	*Tricholoma lascivum*				1	II		
M	*Strobilomyces strobilaceus*				1	II		
M	*Russula alutacea*				2	III		
S	*Otidea onotica*				2	II		
M	*Boletus erythropus*				1	III		
M	*Russula amoena*				1	I	1	

Table II. (continued)

CLASS		ALNETEA	QUERCO-FAGETEA					QRP
ORDER		Aln	Fagetalia				Qpp	Qrp
ALLIANCE Number of associations analized		Alg 2	A-P 8	T-A 2	Car 2	Fag 10	Qpp 2	Qrp 3
M	*Cortinarius hinnuleus*				1	I	1	
M	*Leccinum griseum*				2		1	
M	*Hygrophorus lindtneri*				2	II	2	
M	*Lactarius cremor*				1	II	1	
M	*Russula rosea*				2	V	1	
M	*Russula virescens*				2	IV	2	
M	*Russula foetens*				2	IV	3	
S	*Ramaria formosa*				1	II	1	
M	*Lactarius uvidus*				1	I	1	
M	*Russula aurata*				2	II	2	
M	*Hygrophorus chrysodon*				1	II	1	
M	*Boletus luridus*				1	II	2	
M	*Pulveroboletus gentilis*					I	1	
M	*Hygrophorus penarius*					I	2	
M	*Boletus satanas*					I	2	
M	*Boletus appendiculatus*					I	1	
M	*Russula maculata*					I	1	
M	*Lactarius glaucescens*					I	1	
M	*Lactarius fuliginosus*					I	1	
M	*Lactarius decipiens*					I	1	
M	*Russula grisea*					I	1	
M	*Russula heterophylla*					I	1	

Table II. (continued)

CLASS		ALNETEA	QUERCO-FAGETEA					QRP
ORDER		Aln	Fagetalia				Qpp	Qrp
ALLIANCE		Alg	A-P	T-A	Car	Fag	Qpp	Qrp
Number of associations analized		2	8	2	2	10	2	3
M	*Russula luteotacta*					I	1	
M	*Cortinarius vibratilis*					I	1	
S	*Geastrum fimbriatum*					I	1	
M	*Russula mairei*		1			IV		
M	*Porphyrellus porphyrosporus*					I		
M	*Phylloporus pelletieri*					I		
M	*Gomphus clavatus*					I		
M	*Sarcosphaera eximia*					I		
M	*Boletus regius*						2	
M	*Tricholoma acerbum*						2	
M	*Boletus aereus*						1	
M	*Tricholoma orirubens*						1	
M	*Cortinarius duracinus*						2	
M	*Hygrophorus russula*						2	
M	*Cortinarius rufoolivaceus*						2	
M	*Cortinarius cotoneus*						2	
M	*Amanita caesarea*						1	
S	*Volvariella pusilla*						1	
S	*Bovista plumbea*						1	
M	*Melanogaster variegatus*						1	

Table II. (continued)

CLASS		ALNETEA	QUERCO-FAGETEA					QRP
ORDER		Aln	Fagetalia				Qpp	Qrp
ALLIANCE		Alg	A-P	T-A	Car	Fag	Qpp	Qrp
Number of associations analized		2	8	2	2	10	2	3
Part C								
S	*Lepiota castanea*		I	1		I	2	
S	*Entoloma juncinum*		II		2	I	1	
S	*Laccaria tortilis*		III		1	I		
S	*Mitrophora semilibera*		II	1	1		1	
S	*Tarzetta cupularis*		III		2	II		
S	*Morchella esculenta*		I		2		1	
S	*Calocybe gambosa*		I		2		1	
S	*Agrocybe praecox*		I		2	I	2	
S	*Clitocybe odora*		I		2	II	2	
M	*Inocybe asterospora*		I	1	2	II	2	
S	*Clitocybe gibba*		II	1	2	III	3	
M	*Craterellus cornucopioides*			1	2	IV	2	
S	*Lycoperdon echinatum*			1	2	III	1	
M	*Cortinarius torvus*			1		II	2	
S	*Clavariadelphus pistillaris*			1		II	2	
M	*Inocybe corydalina*			1	2	I	2	
S	*Clitopilus prunulus*			1	2	I	3	
S	*Calvatia excipuliformis*			1			1	
M	*Lactarius ichoratus*			1	1	I	2	

Explanations:
L.f. = life form : M = mycorrhizal (based mainly on Trappe 1962)
P = parasitic
S = saprophytic (on humus and litter)
1, III . . . = constancy in the alliance
Aln = Alnetalia glutinosae; Alg = Alnion glutinosae and A-P = Alno-Padion = compare Table I
T-A = Tilio-Acerion: Aceri-Fraxinetum Koch 1926: Darimont (1973); Ulmo glabrae-Tilietum K.-Lund ap. Seib. 1969: Markussen (unpubl.); Car = Carpinion: Galio silvatici-Carpinetum Oberd. 1957: Lisiewska (1965); Tilio-Carpinetum Tracz. 1962: Ławrynowicz (1973), Nespiak (1959), Wojewoda (1975); Fag = Fagion sylvaticae: Abieti-Fagetum serbicum Jov. 1955: Lisiewska, Jelić (1971), Tortić, Lisiewska (1972); Carici-Fagetum Moor 1952: Jahn *et al.* (1967); Dentario enneaphyllidis-Fagetum (Preis 1938) Oberd. 1957: Gumińska (1962); Dentario glandulosae-Fagetum Klika 1927 em. Mat. 1964: Domański *et al.* (1960, 1963, 1967), Gumińska (1962), Lisiewska (1974), Sałata (1972); Fagetum montanum serbicum Kank. et Miš.: Lisiewska, Jelić (1971); Fagetum praealpinum Leischner-Siska (1939); Luzulo-Fagetum Meusel 1937: Domański *et al.* (1960), Gumińska (1962), Tortić, Lisiewska (1972); Melico-Fagetum Lohm. ap. Seib.: Lisiewska (1963, 1966, 1974); Mercuriali-Fagetum F. Celiński 1962: Lisiewska (1963); Musco-Fagetum Jov.: Lisiewska, Jelić (1971); Qpp = Quercion petraeo pubescentis: Lithospermo-Quercetum Br.B1.1932: Darimont (1973); Potentillo albae-Quercetum Libb. 1933: Ławrynowicz (1973); Potentillo-Quercetum pannonicum Klika: Šmarda (1972); Qrp = Quercion robori petraeae: Fago-Quercetum petraeae Tx. 1955: Bujakiewicz (1987a), Lisiewska (1963, 1966); Querco (roboris) — Betuletum Tx. 1937: Jansen (1981); Violo-Quercetum Oberd. 1957: Jansen (1981).

represented mainly by fleshy mycorrhizal species: *Russula versicolor, Paxillus involutus* and *Inocybe lanuginosa*.

The group of humicolous and litter saprotrophs is represented mostly by *Conocybe, Coprinus, Mycena* and *Psathyrella*. They are diversified in species but occur solitarily.

Salix associates are more pronounced in the Salicetum pentandro-cinereae, e.g. *Naucoria alnetorum, N. langei* and *Hebeloma pusillum*. Bon and Géhu (1973) and Winterhoff (pers. comm.) have given more examples of alder (*Gyrodon lividus*) and willow (*Inocybe salicis, Naucoria salicis*) biotrophs.

3.3. Alno-Padion Knapp 1942 em. Medw. -Korn. ap. Mat. et Bor. 1957

All azonal forests of this alliance occur on moist and fertile soils, mainly on black earth type and on alluvial soil. The breakdown of litter is quick and forms typical mull soil conditions.

The most characteristic feature of the riverside and floodplain forests is the occurrence of few but significant humicolous saprophytic fungi confined to rich, fertile soil. They demand high humidity (hygrophytes) and form small, delicate ephemeral fruit-bodies on black humus, such as *Conocybe mairei, Coprinus cortinatus* and *Laccaria tortilis*. The most typical species of the alliance are: *Entoloma strigosissimum, Mitrophora semilibera* and *Psathyrella vernalis*. Interesting is the occurrence of *Melanophyllum haematospermum*, recognized also as a differential species of the Alno-Padion in the Rhine valley (Carbiener 1981).

The riverside and floodplain forests undergo seasonal floodings which cause a disturbance by strong river action and an enrichment of the soil in nitrogen substances. Thus many weeds are found in the luxuriant field layer. A similar reaction is observed in the mycoflora. Flood sediments, rich in nitrogen compounds, bring about the occurrence of peculiar representatives of *Conocybe*, forming tiny fruit-bodies: *Conocybe arrheni, C. filaris, C. vestita* (Bujakiewicz 1985, 1989). Also *Lepiota cristata, Agrocybe erebia* and *Entoloma icterinum* often grow in such habitats. Their occurrence is greatly favoured by rich substrate (leaf-mull, ruderal places, gardens) and high humidity.

On fertile alluvial soils very few treestand dominants form ectomycorrhiza. Neither *Fraxinus* nor *Ulmus* have associates of that type, whereas *Alnus* and *Salix* do. Alder associates are usually more diversified in species in the Alno-Padion than in the Alnion glutinosae, e.g. *Cortinarius bibulus, C. alnetorum* and *Lactarius obscuratus* have a distinctly higher frequency and abundance. According to Bas (pers. comm.) two alder associates, *Lactarius lilacinus* and

Russula pumila, favour more fertile soils with mobile ground water. Rarely if ever is the whole richness of species of genera mycorrhizal with *Alnus* represented in a given area. In some, *Naucoria subconspersa* is common, in others it does not occur at all. The same holds for *Cortinarius bibulus* and *C. helvelloides, Lactarius obscuratus* and *L. omphaliformis*. It is probably due to a specific combination of edaphic and climatic demands of symbiotic fungi and of their interrelations with the host tree and their special geographic occurrence.

An analysis of mycoflora of particular forest associations of the Alno-Padion (Table I) proves that many have exclusive species and fungi connecting the Alno-Padion, and the Alnion glutinosae form a very distinct group. All five species recognized by Fellner (1980) as characteristic of the Alnion and the Alno-Padion are in this group: *Cortinarius alnetorum, C. helvelloides, Lactarius lilacinus, L. obscuratus* and *Naucoria escharoides*. Interesting is the correlation between fungi and habitats differing in humidity. Fungi listed for the Violo-Ulmetum show that they can live on a habitat with a high degree of insolation. In the Alno-Prunetum in Norway a fairly numerous section of species form components of mixed or pure coniferous stands such as: *Entoloma conferendum, Limacella guttata* and *Rhodocybe nitellina*.

The majority of litter saprotrophs is common with the Alnion glutinosae. Interesting is the group confined to the Alno-Padion, consisting, among others, of *Simocybe centunculus* and *S. rubi*. Also *Mycena speirea* and *Marasmius epiphyllus*, growing on tiny twigs and leaf petioles, seem to be preferred by the Alno-Padion.

Several seasonal fungus aspects are recognized in riverside forests: an early spring aspect with *Mitrophora semilibera* and *Psathyrella vernalis* in the Ficario-Ulmetum campestris and a late summer and autumn aspect of alder associates especially distinct in the Fraxino-Alnetum (Bujakiewicz 1973). In the eastern part of North American floodplain forests *Microstoma floccosum* appears in the late spring (and early summer) and *Helvella sulcata* in the late summer and early autumn (Bujakiewicz 1985).

3.4. Tilio-Acerion Klika 1955

Forests of this alliance occur on warm locations, usually on south-west facing hillsides on rich, rather dry soil. Mycologically they merge into the Alno-Padion (*Entoloma icterinum, Inocybe calospora*) and into the Carpinion and Fagion sylvaticae (*Agrocybe erebia, Inocybe corydalina, I. maculata*). Darimont's (1973) suggestion about the Lepiotetum bucknallii mycosociety seems to be confirmed by the occurrence of *Cystolepiota bucknallii* and *C.*

hetieri and of many hygrophilous and nitrophilous species common with the Alno-Padion.

Locations rich in lime nourish certain calciphilous species, e.g. *Lactarius ichoratus*. In the Ulmo-Tilietum in Norway *Camarophyllus pratensis* and *Hygrocybe coccinea* have a high frequency and abundance.

Though the paucity of data does not allow us to draw detailed conclusions, worth noticing is the group of fungi representing the broad ecological range (Table II, part C) especially those avoiding the Tilio-Acerion.

3.5. Carpinion Oberd. 1953

The zonal forests of the Carpinion occur mainly in the lowlands on fertile brown soils and show distinct aspects in the luxuriant field layer.

Fungi common with the alluvial forests confirm to the richness of habitat, e.g. *Melanophyllum haematospermum* (Lange 1974), *Inocybe fastigiata* and *Entoloma juncinum*.

Leaf shedding brings an abundant litter to decompose. Mycelium of many saprotrophs overgrows and sticks leafy litter (*Clitocybe odora*) and forms large fleshy fruit-bodies, often in 'fairy rings', as *Lepista nebularis* and *L. nuda*. Fungi indicative of well-developed mull conditions (*Clitocybe gibba, Craterellus cornucopioides, Russula lutea*) are especially abundant here.

Fungi occurring in the oak-hornbeam forests in large numbers are associated with *Quercus* (*Lactarius quietus, L. chrysorrheus*), *Carpinus* (*Leccinum griseum, Lactarius pyrogalus*) and *Corylus* (*Hygrophorus lindtneri*). The constant *Carpinus* associates account for the name of the fungus association Leccino (grisei) — Lactarietum circellatae (Šmarda 1972).

The common share of *Fagus sylvatica* in the oak-hornbeam forests brings many associates along (*Hygrophorus eburneus, Russula fellea*). *Amanita phalloides*, an associate of *Quercus* and *Fagus* is usually considered to find his most favourable conditions in the oak-hornbeam forests. Bon and Géhu (1973) and Carbiener *et al.* (1975) also estimated it as locally different from the Carpinion. This fungus, however, shows a very broad amplitude and grows even in mixed and coniferous forests within the limits of *Quercus* (Lange 1974).

Seasonal aspects of the field layer have simultaneous and overlaping series of fungus aspects, formed, among others, with *Sclerotinia tuberosa*, parasitic on rhizomes of *Anemone nemorosa* in very early spring, various species of *Morchella* in early spring, *Calocybe gambosa* in spring, and with *Lepista nuda, L. nebularis* and other saprophytes in autumn (Lisiewska 1965).

3.6. Fagion sylvaticae Tx. et Diem. 1936

Beech forests are generally rich in terrestrial fungi. The synthesis on participation of macrofungi in beech forests, even within the eastern part of the *Fagus* area only (Lisiewska 1972, 1974), gives a convincing picture of their connections with various beech forest types and of the abundance and diversity of mycoflora in this habitat.

The fungus flora of the lowland and morainic Melico-Fagetum seems to be very similar to the oak-hornbeam forest (Nespiak 1968). They generally have the same type of organic matter breakdown and only *Fagus sylvatica* is different by bringing many associates. *Fagus* is a host tree that forms a great variety of mycorrhizal associations, both in the fertile beech forests (*Russula alutacea, R. cyanoxantha, R. fellea*) and in the poor beech woods with raw humus (mor) of the Luzulo-Fagetum type (*Ramaria formosa, Lactarius subdulcis, Russula rosea*).

An abundant accumulation of leaf litter in local depressions or on leeward slopes of ravines brings about the appearance of some groupings of litter saprophytes. They were observed by Jahn *et al.* (1967) and are called 'Clitocybetum'. Mostly *Clitocybe, Collybia* and *Marasmius* are represented in this peculiar assemblage of fungi.

Climatic, orographic, attitudinal and edaphic differences in beech forests are well reflected in the mycoflora.

Theromophilous beech forests, rich in lime, representing the Cephalanthero-Fagion, can be distinguished by the occurrence of *Boletus satanas, Gomphus clavatus, Hygrophorus chrysodon, Russula maculata* and *Sarcosphaera eximia* (Table II, dotted line). Dörfelt (1974) distinguished a separate *Boletus satanas-Boletus radicans* mycocoenosis in the beech forests on lime (after Kreisel 1981).

Fungi indicative of fertile soil often occur in the most humid and rich habitats of the beech forests, namely in the Mercuriali-Fagetum, e.g. *Conocybe mairei, Tubaria pellucida* and even *Naucoria scolecina*.

The montane beech forests (Dentario glandulosae-Fagetum) are distinguished by fungi of montane distribution type, such as *Porphyrellus porphyrosporus* and by many others, such as *Phylloporus pelletieri* and *Cortinarius torvus*. Many fungi of the montane beech forests are common with fir and spruce that stands adjoined to them in the field (*Hygrophorus olivaceoalbus, Lactarius fuliginosus, Cystoderma carcharias*).

A noteworthy assemblage of certain bryophilous fungi was noted in the beech forests (Jahn *et al.* 1967) in places where stemflow water makes the soil more acid than in the forest as a whole. Even some mycorrhizal fungi (*Russula ochroleuca*) prefer these places (Tyler 1984).

The richest in macrofungi are beech forests characterized by slower decomposition of organic matter, e.g. the Luzulo-Fagetum leucobryetosum (Jahn *et al.* 1967). Tyler (1985) found distinct dissimilarities in fungi species composition between beech forest soils of different acidity and organic properties.

Gumińska (1962) studied the rhythm in the occurrence and the seasonal changes of macrofungi in the Dentario glandulosae-Fagetum in southern Poland and distinguished two main aspects: in summer composed by Boletaceae and in autumn consisting of Russulaceae (and *Amanita vaginata*).

3.7. Quercion petraeo pubescentis Jackucs 1961 emend. Medw. Korn. 1972

These are thermophilous oak forests in southern Europe and extrazonal in Central Europe. They grow on dry and shallow alkaline soil rather rich in lime and are often grassy.

Calciphilous fungi are indicative for these habitats, e.g. *Boletus aereus, B. regius, Hygrophorus penarius* and *Tricholoma acerbum. Hygrophorus penarius* is also suggested by Bon and Géhu (1973) and Carbiener *et al.* (1975) as indicative for the Quercion pubescentis. Edaphic and climatic conditions also justify the formation of the Boleto (aerei)-Russuletum luteotactae association (Šmarda 1972). *Boletus satanas* grows in dry oak forests more often than in Central European beech forests. The most characteristic species of the South European calciphilous dry oak woods are: *Amanita caesarea, Boletus appendiculatus* and *Hygrophorus russula* (Table II, dotted line).

Amanita caesarea occurs also in the forests with *Castanea sativa*. According to Lange (1974) this fungus has more mycorrhiza symbionts, as *Pinus* and *Fagus* and grows on the lowlands, mainly around the Mediterranean and southern parts of the Central European mountains. It grows as well in North America, mostly in southern open pine-oak woods but has recently changed the specific epithet to *A. umbonata* Pomerleau (Smith Weber and Smith 1985).

Clitopilus prunulus is fairly numerous on dry oak woods and *Crinipellis stipitaria* often grows on dry grass blades. Open and sunny habitats allow even some xerophilous fungi to grow, e.g. *Bovista plumbea* that occurs on dry pastures (Gross *et al.* 1980) especially in the Lolio-Cynosuretum (Kreisel 1987).

3.8. Quercion robori petraeae Br. Bl. 1932

Forests of this alliance are geographically confined to the western part of Europe and comprise widely distributed Querco (roboris)-Betuletum and two closely related forests — the Violo-Quercetum and the Fago-Quercetum. The soil is acidic and covered with an abundant moss layer.

Fungi distinguishing this syntaxon are typical for coniferous forests (*Clitocybe diatreta, Russula emetica, Tephrocybe tylicolor*) and only the presence of *Quercus* and *Fagus* with their symbionts makes the mycoflora different. Peculiar is the lack of some mycorrhiza formers with *Betula* in the Querco-Betuletum (Jansen 1981). Raw humus is formed and typical representatives of this habitat are common and abundant: *Russula emetica, R. ochroleuca* and *Lactarius camphoratus*. The latter one commonly occurs in the Luzulo-Fagetum.

An abundant moss layer nourishes many bryophilous and saprophytic species of *Galerina* e.g. *G. calyptrata, G. sahleri, G. mniophila, G. pumila.* Wet habitats in this forest type are occupied by *Laccaria proxima*, typical for the Vaccinio-Piceion stands (Kreisel 1987).

3.9. Conclusions

Terrestrial macrofungi, however widely distributed in European forests are rarely if ever randomly distributed. They usually have a broad ecological spectrum, caused mostly by specific demands for a certain combination of edaphic and climatic factors. Some fungi show different ecology in different parts of a given area (*Boletus satanas*). To certain species climatic factors are decisive for their distribution (*Amanita caesarea*), for others edaphic preferences are conclusive (*Melanophyllum haematospermum*).

A comparison and an analysis of mycocoenological data taken from 30 European broad-leaved forest associations (Table II) shows the majority of fungi confined to the Querco-Fagetea class. It reflects general features of the Central European deciduous forests which is paucity in tree genera and emphasizes strong mycorrhizal connections with the few dominant trees over the large areas, especially with *Fagus, Quercus* and *Carpinus.*

Symbiotic fungi are essentially associated with the growth of host trees and the mycorrhizal connections are usually much stronger than the habitat conditions. Saprophytic fungi seem to be more linked with the community type (Table I, part A, C, Table II, part B, C). Both groups of terrestrial fungi show different modes of behaviour especially in seasonal appearance. Vernal and autumnal aspects have usually saprophytic fungi in dominance, whereas mycorrhizal fungi are almost entirely absent.

Most terrestrial fungi occurring in deciduous forests prefer a low degree of insolation and indicate well-developed mull conditions. Distinct seasonal series of fungus aspects are especially well recognized in the forests of the Alno-Padion, Carpinion and Fagion sylvaticae.

Extremely different habitats of the Alnion glutinosae and the Quercion petraeo pubescentis show distinct groups of faithful fungus associates (Table II).

Wet sites of the Alnion glutinosae are characterized by mosaic patterns of habitats, elevated hummocks and boggy hollows, occupied by parallel groups of green plants and fungi.

Among the forest types of the Fagetalia sylvaticae only the Alno-Padion and the Fagion sylvaticae have fungi restricted to a certain syntaxon.

The Alno-Padion is distinguished by saprophytic fungi indicative of fertile and humid habitats, mostly of scattered occurrence. Assemblages of the mycorrhiza formers with *Alnus* are more distinct than in the Alnion glutinosae. Certain saprophytic fungi seem to prefer particular forest associations or groups of forests (Table I).

The Tilio-Acerion seems to merge mycologically with adjoining beech and oak forests.

The majority of fungi growing in the oak-hornbeam forests (Carpinion) is common with various types of mesophytic woods. Considering saprophytic fungi the oak-hornbeam forests are mycologically hardly distinguishable from the lowland beech forests of the Melico-Fagetum. In fact only *Fagus sylvatica* differs essentially in these habitats and makes the picture obscure. *Carpinus* is rather weak in forming ectomycorrhiza and *Quercus*, however strong in this respect, has a much wider ecological spectrum.

Beech forests of the Fagion sylvaticae are developed over a wide range of soils and reflect distinct and many-sided differences in ecological behaviour of fungi species in various parts of the geographic range of these forests. Fertile beech forests growing on brown soils nourishes different species of fungi on lowlands (*Coprinus picaceus*) and in the mountains (*Phylloporus pelletieri*). The frequency of mycorrhiza formers with *Fagus sylvatica* generally increases towards the most acid soils with more or less well-developed mor properties, e.g. in the Luzulo-Fagetum.

Oak woods of the Quercion petraeo pubescentis occupy habitats unfavourable for beech and oak-hornbeam forests. Their proximity makes the mycoflora very similar. Fungi preferential to calcareous soils and many thermophilous and xerophilous species demonstrate this. Both groups include certain taxa forming drought resistant fruit-bodies (*Bovista plumbea, Melanogaster variegatus*).

Acidic and oligotrophic habitats of the Quercion robori petraeae are poor in fungi typical of broad-leaved forests. They seem to occur in areas where coniferous stands would grow if the general climatic conditions would permit it. The presence of raw humus (mor) brings many fungi species in common with coniferous forests.

References

Agerer, R. 1985. Zur Ökologie der Mycorrhizapilze. Bibl. Myc. 97: 1—160.

Agerer, R. 1987. Studies on ectomycorrhizae IX. Mycorrhizae formed by *Tricholoma sulfureum* and *T. vaccinum* on spruce. Mycotaxon 28: 327—361.

Arnolds, E. J. M. 1981. Ecology and coenology of macrofungi in grasslands and moist heathlands in Drenthe, The Netherlands. Part 1. Introduction and Synecology. Bibl. Mycol. 83.

Barkman, J. J. 1973. Synusial approaches to classification. p. 437—491. *In*: R. H. Whittaker (ed.) Handbook of Vegetation Science. Part V. Ordination and Classification. Dr. W. Junk Publishers, The Hague.

Bohus, G., and M. Babos. 1960. Coenology of terricolous macrofungi of deciduous forests. Bot. Jahrb. 80: 1—100.

Bohus, G., and M. Babos. 1967. Mycocoenological investigation of acidophilous deciduous forests in Hungary. Bot. Jahrb. 87: 304—360.

Bohus, G. 1984. Studies on the pH requirement of soil-inhabiting mushrooms: the R-spectra of mushroom assemblages in deciduous forest communities. Acta Bot. Hung. 30: 155—171.

Bon, M., and J. M. Géhu. 1973. Unités superieures de végetation et récoltes mycologiques. Docum. mycol. 6: 1—40.

Bujakiewicz, A. 1969. Udział grzybów wyższych w lasach łęgowych i olesach Puszczy Bukowej pod Szczecinem (Higher fungi in the alluvial and alder forests of the Beech Forest near Szczecin). Bad. Fizjogr. n. Polską Zach. 23(B): 61—69.

Bujakiewicz, A. 1973. Udział grzybów wyższych w lasach łęgowych i w olesach Wielkopolski (Higher fungi in the alluvial and alder forests of Wielkopolska Province). Prace. Kom. Biol. PTPN, 25: 1—91.

Bujakiewicz, A. 1977. Occurrence of macromycetes in floodplain forests along the Marais des Cygnes River. Kansas. USA. Fragm. Flor. et Geob. 23: 87—105.

Bujakiewicz, A. 1981. Grzyby Babiej Góry. II. Wartość wskaźnikowa macromycetes w zespołach leśnych (Fungi of Mt. Babia. II. Indicative value of macromycetes in the forest associations). Acta Mycol. 17: 63—125.

Bujakiewicz, A. in press. Badania mikosocjologiczne w zespole Ficario-Ulmetum campestris Knapp 1942 em. J. Mat. 1976 w rezerwacie Wielka Kępa Ostromecka nad Wisłą (Mycosociological research in the Ficario-Ulmetum campestris Knapp 1942 em. J. Mat. 1976 in the Wielka Kępa Ostromecka reserve on Vistula) Acta Mycol.

Bujakiewicz, A. 1985 (1987). Macromycetes occurring in floodplain forests near Ithaca. New York. USA. Acta Mycol. 21: 165—192.

Bujakiewicz, A. 1986a. Udział macromycetes w zbiorowiskach roślinnych występujących na podłożu torfowym w Słowińskim Parku Narodowym (Occurrence of macromycetes in the plant communities on peaty ground in the Słowiński National Park). Bad. Fizjogr. n. Polską Zach. 37: 101—129.

Bujakiewicz, A. Occurrence of macrofungi in alder carrs on Seili Island. Braun-Blanquetia (in press).

Bujakiewicz, A. 1989. Macrofungi in alder and alluvial forests in various parts of Europe and North America. Opera Botanica 100: 29—41.

Carbiener, R. 1981. Der Beitrag der Hutpilze zur soziologischen und synökologischen Gliederung von Auen- und Feuchtwäldern. Ein Beispiel aus der Oberrheineben. p. 497—531. *In*: Berichte der Intern. Symposien der Intern. Vereinigm für Vegetationskunde. Syntaxonomie. Rinteln.

Carbiener, R., N. Ourisson, and A. Bernard. 1975. Erfahrungen über die Beziehungen zwischen Grosspilzen und Pflanzengesellschaften in der Rheinebene und der Vogesen. Beitr. naturk. Forsch. Südw. Dtl. 34: 37—56.

Darimont, F. 1973. Recherches mycologiques dan les forêts de Haute Belgique. Mém. Inst. Roy. Sc. Nat. Belg. 170.

Domański, S., B. Gumińska, M. Lisiewska, A. Nespiak, A. Skirgiełło, and W. Truszkowska. 1960. Mikoflora Bieszczadów Zachodnich (Wetlina 1958) (Mycoflora of West Bieszczady [Wetina 1958]). Monogr. Botan. 10: 159—237.

Domański, S., B. Gumińska,, M. Lisiewska, A. Nespiak, A. Skirgiełło, and W. Truszkowska. 1963. Mikoflora Bieszczadów Zachodnich. II. (Mycoflora of West Bieszczady. II) Ustrzyki Górne 1960. Monogr. Botan. 15: 3—75.

Domański, S., B. Gumińska,, M. Lisiewska, A. Nespiak, A. Skirgiełło, and W. Truszkowska. 1967. Mikoflora Bieszczadów Zachodnich. III. (Mycoflora of West Bieszczady. III) Baligród 1962. Acta Mycol. 3: 63—114.

Fellner, R. 1980. *Russula pumila* found in Czechoslovakia (with some notes about its distribution and sociology). Česka Mykol. 34: 35—44.

Fellner, R. 1988. Notes to mycocoenological syntaxonomy. 2. The survey of the syntaxonomic classification of mycocoenoses taking into account the principle of the substratum and trophism. Česká mykol. 42: 41—51.

Friedrich, S. 1985/1987. Charakterystyka ekologiczno-fenologiczna macromycetes Puszczy Goleniowskiej (Ecological and phenological characteristic of macromycetes of the Goleniów Forests). Acta Mycol. 21: 143—164.

Gross, G., A. Runge, and W. Winterhoff. 1980. Bauchpilze (Gasteromycetes s.l.) in der Bundesrepublik and Westberlin. Beih. zur Zeitschr. f. Mykologie 2: 1—220.

Gumińska, B. 1962. Mikoflora lasów bukowych Rabsztyna i Maciejowej (Studium florystyczno-ekologiczne) (The fungi of the beech forests of Rabsztyn and Maciejowa) (Floristic-ecological study). Monogr. Bot. 13: 3—85.

Haas, H. 1932. Die bodenbewohnenden Grosspilze in den Waldformationen einiger Gebiete von Württemberg. Beih. Bot. Centralbl. 50B: 35—134.

Jahn, H., A. Nespiak, and R. Tüxen. 1967. Pilzsoziologische Untersuchungen in Buchenwäldern des Wesergebirges. Mitt. Flor.-soz. Arbeitsgem. N.F. 11/12: 159—197.

Jansen, A. E. 1981. The vegetation and macrofungi of acid oakwoods in the North-East Netherlands. Agricult. University Wageningen. Sect. Biological Station Wijster.

Kalamees, K. 1971. Theoretical aspects of mycocoenology. Eesti NSV Teaduste Akad. Toimet. 20 köide. Biologia 2: 150—158.

Kalamees, K. 1979. The role of fungal groupings in the structure of ecosystems. Eesti NSV Teaduste Akad. Toimet. 28 köide. Biologia 3: 206—213.

Kreisel, H. 1981. Pilzsoziologie. p. 62—94. *In*: E. Michael, B. Hennig and H. Kreisel (eds.) Handbuch für Pilzfreunde. Vol. IV. (2nd. ed.) VEB G. Fischer, Jena.

Kreisel, H. 1987. Pilzflora der Deutschen Demokratischen Republik. Basidiomycetes. VEB G. Fischer Verl., Jena.

Lange, L. 1974. The distribution of macromycetes in Europe. Dansk Bot. Arkiv. 30: 1—105.

Ławrynowicz, M. 1973. Grzyby wyższe macroskopowe w grądach Polski Środkowej (Higher macroscopic fungi in the oak-hornbeam forests of the central Poland). Acta Mycol. 9: 133—204.

Leischner-Siska, E. 1939. Zur Soziologie und Oekologie der höheren Pilze. Untersuchung der Pilzvegetation in der Umgebung von Salzburg während des Maximalaspektes 1937. Beih. Bot. Centralbl. 59: 359—429.

Lisiewska, M. 1963. Mikoflora zespołów leśnych Puszczy Bukowej pod Szczecinem (Higher fungi of forest associations of the Beech Forest near Szczecin). Monogr. Botan. 15: 77—151.

Lisiewska, M. 1965. Udział grzybów wyższych w grądach Wielkopolski (Higher fungi of the Querco-Carpinetum of the Wielkopolska province). Acta Mycol. 1: 169—268.

Lisiewska, M. 1966. Grzyby wyższe Wolińskiego Parku Narodowego (Higher fungi in the Wolin National Park). Acta Mycol. 2: 25—77.

Lisiewska, M. 1972. Mycosociological research on macromycetes in beech forest associations. Mycopath. Mycol. appl. 48: 23—34.

Lisiewska, M. 1974. Macromycetes of beech forests within the eastern part of the *Fagus* area in Europe. Acta Mykol. 10: 3—72.

Lisiewska, M. 1978. Macromycetes na tle zespołów leśnych Świętokrzyskiego Parku Narodowego (Macromycetes in forest associations of the Świętokrzyski National Park). Acta Mycol. 14: 163—191.

Lisiewska, M., and M. Jelić. 1971. Mycosociological investigations in the beech forests of some reservations in Serbia (Yugoslavia). Fragm. Flor. et Geob. 17: 147—161.

Nespiak, A. 1959. Studia nad udziałem grzybów kapeluszowych w zespołach leśnych na terenie Białowieskiego Parku Narodowego (The investigations on the character of the correlations between the higher fungi and forest associations in the National Park of Białowieża). Monogr. Bot. 8: 3—141.

Nespiak, A. 1962. Observations sur les champignons à chapeau dan les associations forestières en Pologne. Vegetatio 11: 71—74.

Nespiak, A. 1968. Krytyczne uwagi o socjologii grzybów (Critical remarks on the sociology of fungi). Wiad. Bot. 12: 93—104.

Sałata, B. 1972. Badania nad udziałem grzybów wyższych w lasach bukowych i jodłowych na Roztoczu Środkowym (Studies on the higher fungi in beech and fir forests of the Central Roztocze). Acta Mycol. 8: 69—139.

Smith Weber, N., and A. H. Smith. 1985. A field guide to Southern Mushrooms. The University of Michigan Press, Ann Arbor.

Šmarda, F. 1972. Die Pilzgesellschaften einiger Laubwälder Mährens. Přirodověd Pr. Ustavů ČSAV Brně 6: 1—53.

Tortić, M., and M. Lisiewska. 1972. Mycological investigations in some beech forests of Bosnia (Yugoslavia). Glasnik Zemaljskog Muzelja 10: 65—72.

Trappe, J. M. 1962. Fungus associates of ecototrophic mycorrhizae. Bot. Rev. 28: 538—606.

Tyler, G. 1984. Macrofungi of Swedish beech forest. Dept. of Plant Ecology, Univ. of Lund, Rahms i Lund.

Tyler, G. 1985. Macrofungal flora of Swedish beech forest related to soil organic matter and acidity characteristics. Forest Ecology and Management 10: 13—29.

Winterhoff, W. 1984. Analyses of fungi in plant communities, especially of macro-mycetes. p. 227—370. *In*: R. Knapp (ed.) Sampling methods and taxon analysis in vegetation science. Dr. W. Junk Publishers, The Hague.

Wojewoda, W. 1975. Macromycetes Ojcowskiego parku Narodowego. II. Charakterystyka socjologiczno-ekologiczno-geograficzna (Macromycetes of the Ojców National Park. II. Phytosociological, ecological and geographical characterization). Acta Mycol. 11: 163—209.

4. Macrofungi on soil in coniferous forests

GERHARD KOST

Abstract

A survey is given about species composition of higher fungi in coniferous forest communities. Natural forest communities as well as artificial forest stands are considered. Host range of mycorrhizal fungi, dependence of soil conditions (pH-value, nutrient content, moisture) and climate are discussed in detail.

1. *Artificial spruce forests*

In comparison to natural Piceetum-communities the macromycetes flora of spruce plantation is reduced and varies within a wide range. It consists of frequent and ecologically unspecific species. The dominating fungal species are mostly acidophilic. A complied list of macromycetes species of spruce forests is given. It gives information about abundance, occurrence on different soil types, distribution in different altitudes and areas of Europe. Only some widespread litter decomposing and lignicolous fungi can abundantly be found in spruce plantations. The age of the forest stands distinctly influences the composition of fungal species and their abundance. Disturbances of forest stands caused by man change the natural macromycetes flora. The reduction of species in forest stands is directly dependent on the intensity of forestry management. After manuring with nitrogen or depositing chalk, the fungal flora is transformed by elimination of acidophilic species and by invasion of more neutrophilic fungi.

2. *Fungi of natural Piceetum*

The fungal vegetation of natural spruce forests mainly consists of acidophilic macromycetes. Some neutrophilic species fructificate in Piceetum on chalk-containing soil of deeply weathered limestone. A few species are exclusively distributed in the boreal or subalpine areas of Norway spruce (see list). Most

W. Winterhoff (ed.), Fungi in Vegetation Science, 79—111.

macromycetes associated with Picea abies in natural Piceetum can also be found in spruce plantations in the lowlands.

3. *Fungi in forest communities with silver fir*

Forest communities with silver fir, which grow on sites with good mineral sources and higher soil pH-values, have a characteristic macromycetes flora. Some of these macromycetes can also be associated with broadleaf trees (especially *Fagus* sp.). In silver fir forests on acid soil, some acidophilic fungi commonly found in *Picea*-forest communities also occur.

While soil inhabiting fungi of *Abies* forest communities exactly indicate soil conditions, wood inhabiting fungi of *Abies* are sensitive for climatical conditions.

4. *Fungi in Pinus cembra — Larix decidua forests of central Alps*

Beside ectomycorrhizal species associated with *Pinus cembra* or *Larix decidua*, some other macromycetes, adapted to other species of the genus *Pinus*, can be found in these forest communities. Lists of characteristic species are given.

The litter decomposing and soil-inhabiting Basidiomycetes mostly are unspecific, widespread, and common species of coniferous forests.

5. *Fungi in pine forest communities*

Many ectomycorrhizal and saprophytical macromycetes living in pine forests are stenecously adapted to these forest communities. In many cases the ecological amplitude of *Pinus sylvestris* is broader than that of the associated fungi. Therefore, the fungal vegetation of certain pine forests characterizes the different pine forest communities. Lists of important and characteristic species of several pine forest communities (pine forest on dunes, moor pine forests, *Empertrum-Pinus*-forests, *Pinus sylvestris* forests on chalky soil) are presented. The fungal species composition of pine plantations is discussed.

In sandy (silicatic) pine forests a rich macromycetes flora thrives, but many pine forest associated fungi are strongly threatened. The decline of many macrofungi of pine forests on poor, acid soil is remarkable. But the fungal flora of all indigenous forest communities has to be protected. The problem is that in many cases the knowledge about composition of the fungi in forest communities is deficient. Many additional studies are necessary to elucidate the interactions between fungi and plants in forest communities.

4.1. Introduction

Since the beginning of mycological research, observed environmental conditions were used for characterizing the species (Persoon 1822, 1828, Fries 1821, 1832), However Ferry (1887, 1892) was the first who paid attention to the correlation of fungal flora to their habitat, such as forest type or geological conditions.

It is very difficult to describe the fungal composition of a forest by using one general mycosociological system because of the dissimilar nutritive types of fungi and the various ecology. The different modes of nutrition of fungi lead to the placement of fit theses organisms in quite distinct successional series.

In forest communities of the temperate zone fungi have a big importance characterizing them because of the small number of higher plants here.

4.2. Macromycete species composition of coniferous forests

4.2.1. *Artifical spruce forests*

When communities of fungi are studied and regarded in European forests, one point is both essential and indispensable, and should be kept in mind. The forests in central Europe are geologically young because they have been formed since the end of the last ice age (12 000 years ago) by successive immigration of the tree species into central Europe. Additionally it should be noted that the structure of the forests and their tree composition are altered essentially by the influence of man. After the widespread deforestation in the Middle Ages, many forest stands were planted and recultivated by forestry management since 200—300 years. The most important tree species in forest plantations is Norway spruce (*Picea abies*) and this species is planted from sea level to the subalpine areas. In almost all coniferous mixed forest *Picea abies* was promoted by forestry management in contrast to other tree species.

The overproportional percentage of spruce in almost all coniferous forest communities in Europe has the consequence that the fungi obligatorally associated with spruce and many unspecific ones extended their area and therefore, many fungi are common within the entire central Europe. Gilli (1951) showed this effect in spruce plantations within the area of broad leaf forests (*Fagus sylvatica*).

The roots of spruce release protons into the surrounding soil and therefore the pH-value of the soil is depressed (Marschner 1986). Under these low pH-conditions, the fallen needles form a barely decayable layer of litter decomposition. These influenced soil conditions, the interactions of the fungi with the living trees and the specification of saprophytic substrate decomposition force many similarities in the composition of the fungal flora in spruce stands. However, the mostly poor mycoflora of spruce plantation varies within a wide range.

(a) The dominating group of fungal species are acidophilic, some neutrophilic species can be found in regions with limestone, especially on afforested arable land and old pastures.
(b) The fungal flora is mostly composed of frequent and ecologically unspecific species.
(c) In comparison to natural Piceetum-communities, the species diversity is reduced. Some species are partly advanced and can appear with a high number of fruiting bodies.
(d) Only the common litter decomposing and widespread lignicolous fungi can be found in spruce plantations.
(e) The mountain and subalpine spruce-associated fungi are missing in forest plantations in the lowlands.
(f) Occasionally, disturbances inside forest stands caused by man (road making, deposition of plant debris, manuring) extremely change the natural composition of fungi (Kost 1991a).

Baxter (1937), Bourgeois (1952), Höfler (1955b), Meisel-Jahn and Pirk (1955), Nathorst-Windahl 1956, Haas (1958, 1971, 1975, 1979), Hora (1959, 1972), Ricek (1961), Bujakiewicz (1979, 1981, 1982a, 1982b), Seibt (1981), Agerer and Kottke (1981) and Rücker and Peer (1988) studied macromycetes in forest plantations. Seidel (1983) compared the fungal species composition of young afforestations with that of mature forest stands. The succession of macromycetes which grow in spruce plantation on old pasture lands was recorded (Ricek 1981). With mycorrhizal as well as saprophytic fungi, it can be stated that quite different species fructificate in very young to 30 year old afforestations. Early- and late-mycorrhizal can be distinguished in young and old stands (Thomas *et al.* 1983, Dighton and Mason 1985, Dighton *et al.* 1986, Al Abras *et al.* 1988). In 20 to 40 year old forests stands the number of fungal species is higher than in older forests, where the rate of fructification is also lower.

(a) The mode of forestry management such as thinning, removing wood, use of litter and manuring, disturbs the normal degradation of the litter layer or the usual development of the substrate for fungal succession. There is therefore a strong negative influence on the species composition. Only a reduced number of some unspecific macromycetes are able to survive under these conditions (Winterhoff 1984b,c, Rücker and Peer 1988, Kost and Haas 1989, Kost 1991a). The reduction of species in forest stands is directly dependent on the intensity of forestry management.

Only some unspecific litter decomposers (i.e. *Marasmius androsaceus, Micromphale perforans*, etc.) unspecific mycorrhizal fungi (i.e. *Russula ochroleuca*, etc.) and sometimes parasites (*Heterobasidion annosum, Stereum sanguinolentum*) appear in forest stands in comparatively high abundance (Haas 1975, 1979).

(b) The fructification of macromycetes is reduced and the composition of fungal species is distinctly altered by an increasing growth of herbaceous and shrubby plants on the ground. This can be also caused by several other factors (intensive thinning, road-making, needle loss, forest fertilization, passive deposition of minerals by the atmosphere, compression of the soil surface layers, ground drainage, etc.) (Ohenoja 1978, 1983, 1988a,b). Results of studies in uninfluenced coniferous forests ('Bannwälder') in south-west Germany clearly demonstrates that the number of fungal species is higher there.

After manuring with nitrogen or depositing chalk, the fungal flora is transformed by elimination of acidophilic species (mycorrhizal or saprophytic ones) and by invasion of more neutrophilic fungi (Fiedler and Hunger 1963, Ritter and Tölle 1978, Horak and Röllin 1989, Ohenoja 1988b). Usually only weak and non significant effects after application of phosphorus, potassium, or magnesium have been observed (Kuyper 1988, 1989, Uebel 1982).

Geographical or oreographical differences of the spruce plantation sites are not distinctly manifested, but specific geological and pedeological characteristics are revealed by the mycoflora on sites with the first generation of spruce after afforestation of arable land (Haas 1971, 1975).

The arrangement of litter decomposers and wood rotters in Sitka spruce stands in Scotland is quite similar to artificial Norway spruce stands in other parts of Europe (Alexander and Watling 1987, Alexander 1981). This mycoflora is derived from the native fungal flora of birch-pine forest and most of the fungi are the identical acidophilic, unspecific species (Table I).

Meisel-Jahn and Pirk (1955) observed various groups of species depending on humus conditions and superficial soil acidification in forest plantations. Šmarda (1973) described three fungal associations in secondary spruce forests of Moravia (Czechoslovakia) with the intention of characterizing the spruce forests. In these papers (Meisel-Jahn and Pirk 1955, Svrček 1960, Šmarda 1973) the same ecological factors determine the composition of the fungal flora. Fungal associations described by Šmarda (1973):

Geastro-Agaricetum semotae F. Šmarda	— in spruce stands with a rich flora of vascular plants
Clitocybo-Phellodonetum nigrae F. Šmarda	— in spruce stands with acid soil
Pholioto-Inocybetum acutae F. Šmarda	— in spruce stands with very moist soil.

Studying the occurrence of the parasite *Heterobasidion annosum* in spruce forest stands on the Swabian Mountains, Haas (1979) gave a detailed description of the fungal components. He found a definite correlation between high pH-value (5—7) and a higher amount of neutrophilic macromycetes. Only common needle digesting macromycetes constantly fructificate in these forest stands as characteristic members of coniferous forests.

Compiling the species list given from several spruce forest stands the following macromycetes (Table I) seem to be the most common and characteristic species in Europe:

Table I. Common and widespread macromycetes in spurce forest stands

Russula ochroleuca	*Mycena rosella*
Hygrophorus pustulatus	*Clavulina cristata*
Inocybe umbrina	*Micromphale perforans*
Lactarius necator	*Marasmius androsaceus*
Cystoderma amiantinum	*Clitocybe ditopa*
Mycena galopus	*Clitocybe vibecina*
Mycena pura	*Clitocybe dicolor*

With respect to the great number of soil-inhabiting mycorrhizal or saprophytically-living macromycetes which are associated with *Picea abies*, one can imagine the immense changes in area of distribution of fungi caused by the plantation in central Europe by forestry management or by amplification of the percentage of spruce in forests by selectively cutting the other tree species (*Abies alba, Fagus sylvatica*, etc.) (Table III).

4.2.2. *Fungi of natural Piceetum*

The fungal vegetation of natural spruce forest mainly consists those macromycetes which are adapted to sites with acid soil, raw humus, and sometimes

Table II. Macromycetes of montane or subalpine spruce forests

Amanita pantherina	*Hygrophorus piceae*
Cortinarius speciosissimus	*Lactarius lignyotus*
Cortinarius venetus var. *montanus*	*Lactarius picinus*
Hygrophorus hyacinthinus	*Russula consobrina*
Hygrophorus karstenii	*Russula mustelina*

Table III. Compiled list of macromycetes species of spruce forests

			a_{bm}	b	c	d	e	f	g	h	i	k	l	m	n	o	p
		Agaricus abruptibulbus				d									n		p
		Agaricus arvensis								h			l			o	
		Agaricus haemorrhoidarius													n		
		Agaricus macrocarpus													n		
		Agaricus perrarus															p
		Agaricus rusiophyllus				d											
N		*Agaricus semotus*				d							l				p
N	S	*Agaricus silvaticus*								h		k	l				p
		Agrocybe tabacina															p
		Albatrellus confluens														o	p
		Albatrellus ovinus														o	p
A		*Amanita citrina*							g		i						
A	S	*Amanita fulva*			c												p
CA	S	*Amanita muscaria*	a_b						g	h					n	o	p
A		*Amanita pantherina*								h							p
	S	*Amanita porphyria*														o	
CA	S	*Amanita rubescens*	a_b	b		d	e		g			k				o	p
		Amanita spissa	a_b				e										p
		Amanita submembranacea												m			
C	S	*Amanita vaginata*		b						h	i		l	m		o	p
A	S	*Boletus edulis*	a_b							h					n	o	p
		Boletus luridus														o	
		Calocybe onychina														o	
CA	S	*Cantharellus cibarius*	a_b				e	f						m		o	p
CA	S	*Cantharellus tubaeformis*	a_b				e	f		h		k				o	p
A	S	*Chalciporus piperatus*													n	o	p
		Chroogomphus helveticus															p
		Chroogomphus rutilus														o	
		Clavariadelphus fistulosus														o	
		Clavariadelphus ligula						f		h						o	p
N		*Clavariadelphus pistillaris*				d											p
		Clavariadelphus truncatus														o	
		Clavulina cinerea							g	h				m		o	
C	S	*Clavulina cristata*	a_m	b								k	l	m	n		
		Clitocybe alexandri														o	
		Clitocybe brumalis					e					k					
		Clitocybe candicans														o	
		Clitocybe cerussata													n		
N		*Clitocybe clavipes*	a_b						g	h					n	o	
		Clitocybe dealbata														o	
C	S	*Clitocybe dicolor*					e					k	l				p
		Clitocybe ditopa									i	k	l		n		
C	S	*Clitocybe fragrans*													n	o	
		Clitocybe geotropa													n		p
C	S	*Clitocybe gibba*	a_b	b								k	l		n		p
		Clitocybe gilva						f		h		k	l				
		Clitocybe hydrogramma	a_m									k					p

Table III (continued)

			a_{bm}	b	c	d	e	f	g	h	i	k	l	m	n	o	p
		Clitocybe inversa															p
	S	*Clitocybe metachroa*						f			i						
		Clitocybe odora				d											p
		Clitocybe pityophila								h					n		
		Clitocybe radicellata														o	
		Clitocybe sinopica														o	
		Clitocybe squamulosa				d											
		Clitocybe suaveolens								h		k	l			o	p
		Clitocybe tuba								h							
		Clitocybe vibecina	a_b									k			n		
		Clitopilus cretaceus														o	
		Clitopilus prunulus														o	
C	S	*Collybia asema*	a_m	b	c					h	i	k	l	m	n	o	p
		Collybia cohaerens													n		
		Collybia confluens										k			n		
		Collybia cookei														o	
		Collybia dryophila	a_b										l				p
A	S	*Collybia maculata*							g			k	l				p
		Collybia tuberosa														o	
		Conocybe magnicapitata														o	
		Conocybe tenera														o	
		Coprinus atramentarius														o	
		Coprinus cinereus									i						
		Coprinus comatus														o	
		Coprinus semitalis												m			
		Cortinarius acutus	a_m					f						m			
		Cortinarius allutus	a_b								i						
A		*Cortinarius anomalus*													n	o	p
		Cortinarius armeniacus								h							
		Cortinarius aromaticus															p
		Cortinarius arquatus														o	
		Cortinarius atrocoeruleus		b	c												
		Cortinarius aureofulvus														o	
		Cortinarius bicolor													n	o	
		Cortinarius biformis						f								o	p
		Cortinarius bivelus														o	p
		Cortinarius brunneofulvus															p
CA		*Cortinarius brunneus*	a_b	b	c					h	i			m		o	p
		Cortinarius callisteus														o	p
		Cortinarius calochrous														o	
		Cortinarius camphoratus	a_b				e	f							n		p
		Cortinarius candelaris														o	
M		*Cortinarius caninus*													n	o	
		Cortinarius causticus														o	
C	S	*Cortinarius collinitus*	a_b		c				g			k		m		o	p
		Cortinarius crassus													n		
		Cortinarius decipiens	a_m									k			n		

Table III (continued)

			a_{bm}	b	c	d	e	f	g	h	i	k	l	m	n	o	p
		Cortinarius delibutus													n		p
N		*Cortinarius duracinus*	a_m	b										m		o	
		Cortinarius elegantior														o	
		Cortinarius emollitus														o	
		Cortinarius erythrinus		b													
		Cortinarius evernius					e	f									p
		Cortinarius fasciatus	a_m							h							
	S	*Cortinarius flexipes*													n		p
		Cortinarius fraudulosus														o	
A		*Cortinarius fulvescens*										k					
		Cortinarius fuscoochrascens													n		
A		*Cortinarius gentilis*	a_b									k				o	p
		Cortinarius glandicolor														o	p
		Cortinarius glaucopus													n	o	p
		Cortinarius haematochelis														o	
		Cortinarius herpeticus															p
		Cortinarius illibatus												m			
		Cortinarius infractus	a_m	b													p
		Cortinarius isabellinus									i						
		Cortinarius jubarinus									i						
		Cortinarius junghuhnii		b													
		Cortinarius laniger														o	
		Cortinarius leucopus								h	i						
A		*Cortinarius limonius*						f									
	S	*Cortinarius malachioides*													n		
		Cortinarius microspermus					e	f									
		Cortinarius miniatopus						f									
		Cortinarius mucosus							g	h							
		Cortinarius multiformis								h						o	p
		Cortinarius multivagus															p
CA	S	*Cortinarius obtusus*	a_b				e	f			i						
		Cortinarius ochropallidus									i						
MN		*Cortinarius odorifer*												m			p
CA		*Cortinarius paleaceus*	a_b	b							i			m			p
		Cortinarius papulosus						f									
N		*Cortinarius percomis*														o	
		Cortinarius privignus															p
		Cortinarius pseudosalor									i						
		Cortinarius rigidus									i						
		Cortinarius rubricosus								h							
		Cortinarius rufoolivaceus														o	
		Cortinarius russeoides														o	
		Cortinarius salor							g					m			p
		Cortinarius scutullatus						f									
		Cortinarius sebaceus															p
		Cortinarius sertipes						f									
A	S	*Cortinarius speciosissimus*															

Table III (continued)

			a_{bm}	b	c	d	e	f	g	h	i	k	l	m	n	o	p
		Cortinarius spilomeus													n		p
		Cortinarius subtortus														o	
		Cortinarius subvalidus													n		
		Cortinarius traganus	a_b				e			h						o	p
		Cortinarius umidicola	a_b								i						
		Cortinarius variicolor	a_m											m	n		p
N		*Cortinarius varius*	a_m									k	l				p
		Cortinarius venetus													n		p
		Cortinarius vibratilis				d					i						
N		*Cortinarius vitellinus*														o	
		Cudonia circinans										k				o	p
C	S	*Cystoderma amiantinum*	a_{bm}	b								k	l	m		o	p
CN		*Cystoderma carcharias*	a_m			d		f			i	k	l	m	n	o	p
		Cystoderma granulosum														o	p
		Dermocybe anthracina														o	
C	S	*Dermocybe cinnamomeolutea*	a_b					f				k			n		
		Dermocybe cinnamomea		b						h	i						
		Dermocybe crocea														o	
A		*Dermocybe sanguinea*	a_{bm}					f	g							o	
		Dermocybe semisanguinea	a_b				e	f									
		Disciotis venosa														o	
CA	S	*Entoloma cetratum*	a_b								i	k	l				
		Entoloma clandestinum					e	f									
	S	*Entoloma conferendum*										k		m	n		
		Entoloma cordae									i						
N		*Entoloma nidorosum*					e	f									
		Entoloma papillatum				d											
		Entoloma pascuum							g								
		Entoloma sericeum															p
	S	*Entoloma turbidum*	a_b				e										
		Galerina hypnorum	a_b	b							i						
		Galerina mniophila		b													
		Galerina paludosa			c												
		Galerina rubiginosa															p
		Galerina sahleri		b													
		Galerina sideroides		b													
		Galerina sphagnorum			c												
		Geastrum fimbriatum													n		
		Geastrum pectinatum				d											
		Geastrum quadrifidum															
		Gomphidius glutinosus								h	i	k				o	p
		Hebeloma circinans													n		
		Hebeloma crustuliniforme													n		p
N		*Hebeloma edurum*	a_m									k			n		
		Hebeloma elatum															p
	S	*Hebeloma mesophaeum*													n	o	
		Hebeloma pumilum															p

Table III (continued)

			a_{bm}	b	c	d	e	f	g	h	i	k	l	m	n	o	p
		Hebeloma strophosum															p
		Hebeloma subsaponaceum														o	
		Hemimycena delicatella													n		
		Hemimycena gracilis												m			
		Hemimycena lactea								h							
		Hemimycena pseudocrispula															p
	S	*Hydnum repandum*	a_m							h		k				o	
A	S	*Hydnum rufescens*														o	
CA	S	*Hygrophoropsis aurantiaca*	a_b								i		l				
CN		*Hygrophorus agathosmus*	a_m							h		k	l	m	n	o	p
N		*Hygrophorus chrysodon*	a_m											m			p
CN	S	*Hygrophorus discoideus*	a_m									k	l	m		o	p
		Hygrophorus eburneus															p
		Hygrophorus erubescens	a_m											m		o	p
M		*Hygrophorus hyacinthinus*												m		o	
M		*Hygrophorus karstenii*												m			
A		*Hygrophorus olivaceoalbus*		b	c			f		h		k	l	m		o	
M		*Hygrophorus piceae*	a_m											m		o	
CN	S	*Hygrophorus pustulatus*	a_m	b							i	k	l	m	n	o	
A		*Hygrophorus tephroleucus*															p
A		*Inocybe acuta*	a_m					f								o	
		Inocybe albidodisca										k					
		Inocybe appendiculata													n		
		Inocybe boltonii															p
N		*Inocybe bongardii*												m			p
		Inocybe calamistrata														o	
N		*Inocybe cervicolor*	a_m									k		m	n		p
N		*Inocybe cincinnata*	a_m									k	l	m	n		
		Inocybe commutabilis													n		p
		Inocybe decissa														o	
		Inocybe destricta															p
	S	*Inocybe eutheles*	a_m									k	l				
N		*Inocybe fastigiata*	a_m											m			p
CN		*Inocybe friesii*	a_m									k	l	m	n	o	p
		Inocybe fuscomarginata														o	
CN	S	*Inocybe geophylla*	a_m			d						k	l	m		o	p
		Inocybe grammata															p
		Inocybe hypophaea														o	
		Inocybe lacera														o	
	S	*Inocybe lanuginosa*														o	
		Inocybe leptocystis	a_m									k					
		Inocybe leucoblema										k					p
		Inocybe lucifuga													n		
		Inocybe lutescens														o	
		Inocybe mixtilis											l	m	n	o	
A	S	*Inocybe napipes*	a_b														
		Inocybe oblectabilis														o	p

Table III (continued)

			a_{bm}	b	c	d	e	f	g	h	i	k	l	m	n	o	p
	S	*Inocybe obscura*													n		
		Inocybe ovatocystis															p
		Inocybe paludinella															p
		Inocybe phaeosticta															p
		Inocybe praetervisa													n	o	
		Inocybe pseudohiulca														o	
		Inocybe pudica														o	
		Inocybe pusio														o	
		Inocybe pyriodora															p
N		*Inocybe terrigena*	a_m									k					p
		Inocybe umbratica															p
A	S	*Inocybe umbrina*															p
	S	*Inocybe virgatula*													n		
C	S	*Laccaria amethystea*	a_b	b	c				g		i	k	l	m	n	o	
C	S	*Laccaria laccata*	a_{bm}	b				f	g	h	i	k	l	m	n		p
		Lactarius badiosanguineus	a_m					f						m			p
	S	*Lactarius camphoratus*										k					
N	S	*Lactarius deterrimus*													n		p
		Lactarius fuliginosus															p
A		*Lactarius helvus*	a_b	b	c		e	f	g	h						o	
N		*Lactarius ichoratus*	a_m									k				o	
MA		*Lactarius lignyotus*	a_b	b						h				m		o	p
		Lactarius mammosus	a_b				e	f									
C		*Lactarius mitissimus*	a_m					f	g	h	i	k	l			o	p
CA		*Lactarius necator*	a_b	b	c					h	i	k	l				
MA		*Lactarius picinus*	a_m											m	n	o	
CA	S	*Lactarius rufus*	a_b	b	c	d	e	f		h	i	k	l			o	p
N		*Lactarius scrobiculatus*	a_m									k		m	n	o	p
N		*Lactarius semisanguifluus*	a_m									k				o	
A		*Lactarius sphagneti*						f									
CA	S	*Lactarius theiogalus*	a_b					f	g			k	l				
		Lactarius uvidus															p
		Lactarius vietus			c					h							
		Lactarius zonarioides															p
		Lepiota clypeolaria										k	l				
		Lepiota cristata								h							
		Lepiota ventriosospora				d									n		
		Lepista glaucocana													n		
		Lepista inversa											l		n		
		Lepista irina													n		
		Lepista nebularis								h			l		n		
C	S	*Lepista nuda*				d				h			l		n	o	
		Leucocortinarius bulbiger														o	
		Leucopaxillus alboalutaceus															p
CN		*Limacella glioderma*	a_m			d						k	l				
		Limacella guttata				d											
		Lycoperdon gemmatum								h							

Table III (continued)

			a_{bm}	b	c	d	e	f	g	h	i	k	l	m	n	o	p
		Lycoperdon perlatum													n	o	
		Lycoperdon umbrinum															p
		Lyophyllum infumatum														o	
		Lyophyllum loricatum														o	
		Lyophyllum seminatile														o	
N		*Macrocystidia cucumis*											l				
		Macrolepiota gracilenta														o	
CN		*Macrolepiota procera*	a_m			d						k	l		n		p
		Macrolepiota rhacodes		b					g						n	o	
		Marasmiellus ramealis									i						
C	S	*Marasmius androsaceus*	a_b	b	c						i	k	l	m	n	o	
		Marasmius bulliardii												m	n		
		Marasmius scorodonius													n		
		Marasmius wynnei													n		
		Melanoleuca graminicola														o	
		Melanoleuca vulgaris										k					
N		*Melanophyllum echinatum*														o	
		Melastiza chateri														o	
C	S	*Micromphale perforans*	a_{bm}	b	c					h	i	k	l	m	n	o	
		Morchella conica														o	
		Morchella esculenta														o	
		Mycena acicula														o	
		Mycena aetites									i						p
C		*Mycena alcalina* s.l.		b							i			m			p
		Mycena amicta												m		o	
		Mycena amygdalina									i						
C		*Mycena aurantiomarginata*								h	i			m	n	o	
		Mycena chlorinella											l	m			
C	S	*Mycena epipterygia*	a_m							h	i	k	l	m	n		p
		Mycena epipterygioides															p
		Mycena filopes					e										
		Mycena flavescens													n		
C		*Mycena flavoalba*									i	k	l	m			
C		*Mycena galopus*	a_{bm}	b	c		e	f			i	k	l	m	n	o	p
		Mycena laevigata												m			
		Mycena leptophylla														o	
		Mycena maculata	a_b											m			
		Mycena metata				d				h							
C		*Mycena phyllogena*	a_m								i		l		n		
		Mycena pithya										k					
		Mycena polygramma														o	
C	S	*Mycena pura*	a_m		c	d				h	i		l	m	n	o	p
		Mycena rorida		b	c						i						p
C		*Mycena rosella*	a_m		c			f			i	k	l		n	o	p
C		*Mycena rubromarginata*									i	k	l	m		o	p
C		*Mycena sanguinolenta*									i				n		
		Mycena viridimarginata												m			

Table III (continued)

			a_{bm}	b	c	d	e	f	g	h	i	k	l	m	n	o	p
		Mycena viscosa												m			
	S	*Mycena vitilis*														o	
		Mycena vulgaris												m	n	o	p
		Mycena zephirus													n		
		Mycenella margaritispora														o	
A	S	*Omphalina ericetorum*	a_b		c									m			
		Otidea abietina														o	
		Otidea onotica														o	
		Oudemansiella platyphylla															o
		Panaeolus guttulatus														o	
		Paxillus atrotomentosus	a_b								i		l				
CA	S	*Paxillus involutus*	a_b	b					g	h	i	k				o	p
		Peziza succosa														o	
	S	*Phallus impudicus*				d			g								
		Phellodon niger					e					k					
		Phellodon tomentosus														o	
		Phlogiotis helvelloides														o	
		Pholiota carbonaria														o	
A	S	*Pholiota scamba*	a_b					f									
		Pholiota spumosa															p
		Pholiotina blattaria														o	p
		Pholiotina exannulata														o	
		Pholiotina filaris														o	
		Polyporus leucomelas															p
		Porphyrellus pseudoscaber															p
		Psathyrella albidula														o	
		Psathyrella velutina														o	
		Ramaria flava														o	
		Ramaria ochraceovirens														o	
		Ramaria stricta								h							
N		*Rhodocybe nitellina*				d						k		m			p
		Rickenella fibula							g		i						
		Rickenella swartzii															p
		Ripartites heleomorphus													n		
		Ripartites tricholoma													n		p
A		*Rozites caperatus*	a_b		c		e									o	p
		Russula adulterina														o	
		Russula adusta														o	
		Russula amethystina									i						
		Russula aurata														o	
		Russula badia								h							
		Russula carminea						f									
		Russula cessans									i						
M		*Russula consobrina*			c												p
		Russula cyanoxantha								h		k					p
A		*Russula decolorans*	a_b				e									o	p
		Russula delica													n	o	p

Table III (continued)

			a_{bm}	b	c	d	e	f	g	h	i	k	l	m	n	o	p
CA	S	*Russula emetica*	a_b	b	c					h		k				o	
		Russula emeticella					e	f									
A		*Russula fellea*									i	k	l				
N		*Russula firmula*	a_m									k	l	m			
		Russula foetens														o	p
A		*Russula fragilis*						f		h					n		
C		*Russula integra*	a_m							h		k	l		n		
		Russula lepida				d											
MA		*Russula mustelina*	a_b	b				f								o	p
CA	S	*Russula nauseosa*	a_m								i	k			n	o	p
	S	*Russula nigricans*										k					p
		Russula nitida								h							
CA	S	*Russula ochroleuca*	a_b	b				f		h	i	k	l		n	o	
		Russula olivacea	a_m								i						
		Russula olivascens															p
A		*Russula paludosa*								h							
		Russula pectinata													n		
C	S	*Russula puellaris*						f		h	i	k	l				p
CN		*Russula queletii*	a_m	b	c										n	o	p
N		*Russula torulosa*										k					
		Russula turci															p
		Russula vesca													n	o	p
		Russula veternosa															p
		Russula viscida														o	p
	S	*Russula xerampelina*										k			n	o	p
		Sarcodon cyathiforme															p
	S	*Sarcodon imbricatum*														o	p
		Spathularia flavida														o	
		Stropharia aeruginosa															p
		Tephrocybe inolens										k			n		
N		*Tephrocybe ozes*	a_m									k	l				p
N		*Tephrocybe rancida*											l				p
		Thelephora palmata													n		
		Tricholoma albobrunneum														o	
N		*Tricholoma aurantium*											l				p
		Tricholoma bufonium										k					
		Tricholoma imbricatum														o	
N		*Tricholoma inamoenum*	a_b									k					p
N		*Tricholoma orirubens*														o	
		Tricholoma portentosum														o	
C		*Tricholoma saponaceum*	a_m							h	i	k		m		o	p
		Tricholoma scalpturatum										k					
		Tricholoma stans													n		
		Tricholoma sulphureum											l	m			p
N		*Tricholoma terreum*	a_m									k	l				p
		Tricholoma triste														o	
CN		*Tricholoma vaccinum*	a_m									k			n	o	p

Table III (continued)

			a_{bm}	b	c	d	e	f	g	h	i	k	l	m	n	o	p
		Tricholoma virgatum															p
M	S	*Tricholoma viridilutescens*						f									p
		Tubaria furfuracea														o	
		Volvariella surrecta														o	
CA	S	*Xerocomus badius*	a_b	b			e	f		h		k					p
CA		*Xerocomus chrysenteron*								h	i	k					
		Xerocomus spadiceus														o	
M	S	*Xerocomus subtomentosus*														o	p

a_b = growing on soil of sandstone (Schwöbel 1971)
a_m = growing on soil of limestone (Schwöbel 1971)
a_{bm} = growing on a_b and a_m (Schwöbel 1971)
b = Nespiak 1971
c = Nespiak 1962/63
d = Geastro-Agaricetum semotae (Šmarda 1973)
e = Clitocybo-Phellodonetum nigrae (Šmarda 1973)
f = Pholioto-Inocybetum acutae (Šmarda 1973)
g = Meisel-Jahn and Pirk 1955
h = Höfler 1955
i = Agerer and Kottke 1981
k = Haas 1971
l = Haas 1979
m = Schmid-Heckel 1985
n = Ricek 1981
o = Horak 1963
p = Horak 1985
S = Sitka spruce forest (Alexander and Watling 1987)
C = common species
A = Acidophilic
N = Neutrophilic
M = montane — subalpine

a dense cover of sphagnum. While most macromycetes which are associated with *Picea abies* and naturally grow in Piceetum can also be found in spruce plantations and afforestations in the lowlands, only a few species are exclusively distributed in the boreal or subalpine areas containing Norway spruce.

It is therefore obvious that differences of the composition of macromycetes between artificial and genuine spruce forests exists (Höfler 1955, Favre 1960, Horak 1985, Dörfelt 1981). The mountain and subalpine lignicolous fungi are better suited to characterize a Piceetum subalpinum than the soil inhabiting fungi (Jahn 1969). Favre (1955), Horak (1963, 1985), Jahn (1969), Nespiak (1960, 1971), Engels and Friedrich (1971,

1974, 1976), Kalamees (1980) and Schmid-Heckel (1985) published detailed lists of species from subalpine spruce forests. The fungal species composition of the Piceetum excelsae carpaticum shows good similarities to subalpine Piceeta (Bujakiewicz 1982).

The diversity of fungal species of various ecological groups is prominent in these forests (Nespiak 1971), since quite discrete and small habitats are arranged in a mosaic like manner inside the forest (Höfler 1955b). In addition to acidophilic soil preferring macromycetes of sites with layers of undigested plant debris and with cushions of Sphagnum-plants, some neutrophilic species fructificate on chalky soil of deeply weathered limestone (*Cortinarius odorifer, Lactarius deterrimus, Tricholoma vaccinum, Tr. aurantium,* etc.) (Dörfelt 1981).

4.2.3. *Fungi in forest communities with silver fir*

Forest communities with silver fir (*Abies alba*) are classified by Oberdorfer (1987) within the class Vaccinio-Piceetea Br. -Bl. 1939, suballiance Vaccinio-Abientenion Oberd. 1962, and within the alliance Fagion sylvaticae Pawl. 1928 in the class Querco-Fagetea. In the fungal species composition there are large differences between the two forest communities. The fungal vegetation of Vaccinio-Abietetum, which grows at the mountain level on slightly acidic soil, clearly shows similarities to spruce forests lacking silver fir; it is not easy to distinguish them (Stricker 1950, Haas 1958, 1972, Schwöbel 1971, Haas and Kost 1985, Kost and Haas 1989). In forests on acid soil, certain species of acidophilic macromycetes which are more commonly found in *Picea*-forest communities fructificate (Jahn 1968, Carbiener *et al.* 1975a,b, Lisiewska 1978, Bujakiewicz 1982, Schmid-Heckel 1985). However the macromycetes preferring very acid conditions are lacking in mixed coniferous forests containing silver fir.

Galio-Abietenion Oberd. 1962 and the beech forest of the Eu-Fagion Oberdorfer 1957 with *Abies alba* (Abieti-Fagetum) grow on sites with good mineral sources and higher soil pH-values. Here, most of the macromycetes are associated with deciduous trees or digest plant debris from them. However, some species, especially soil-inhabiting fungi, are very typical for the Galio-Abietetum (Table IV). It is interesting that some species, which normally use *Abies* as partner, are also able to utilize *Fagus sylvatica*. On the other hand fungal partners of coniferous trees or specialists of spruce have no importance in *Abies* communities.

In mixed *Abies alba-Picea abies* forests, which can exist with various soil conditions, Haas (1958) showed pairs of agarics which distinguished the soil pH-value, so that only one of the two species was able to fructificate in the forest (Table V) (Haas 1958).

Since many stenecous fungi specifically react to the mosaicly structured vegetation of the sites with their varying environmental conditions, the summarized checklist of fungi often contains species from quite different ecological places. Since silver fir forests are usually interlaced with other forest communities (Piceetum, Luzulo-Fagetum, Aceri-Fraxinetum . . .), it is sometimes very difficult to distinguish those fungi closely associated to *Abies* from those which belong to other trees or frequently are members of other forest communities (Kotlaba and Pouzar 1962, Carbiener *et al.* 1975a,b, Guminska 1976, 1989, Kost and Haas 1989, Laber and Laber 1989).

Table IV. Soil inhabiting *Abies* associated macromycetes

Obligatory *Abies* partners		Facultative *Abies* partners (sometimes associated with *Fagus*)	
Bondarzewia montana	L	*Cortinarius percomis* pC	M
Caloscypha fulgens pC	M?	*Cortinarius vitellinus* pC	M
Cortinarius atrovirens pC	M	*Hygrophorus capreolarius* pC	M
Cortinarius nanceiensis pC	M	*Hygrophorus chrysodon* pC	M
Ganoderma carnosum	L	*Hygrophorus marzuolus* pC	M
Hygrophorus pudorinus pC	M	*Lactarius ichoratus* pC	M
Inocybe abietis pC	M	*Meripilus giganteus*	L
Lactarius glutinopallens pC	M	*Mycena crocata*	S
Lactarius salmonicolor pC	M	*Mycena vulgaris*	S
Oudemansiella melanotricha pC	L	*Russula amethystina*	M
Podofomes trogii pC	L	*Russula viscida*	M
Russula cavipes pC	M	*Tricholoma orirubens* pC	M

M = ectomycorrhizal fungus
L = lignicolous
S = saprophytic on plant debris
pC = preferring chalky soil

Table V. Pairs of agarics in mixed *Abies alba-Picea abies* forests on different soils

On limestone	–	On sandstone
Cortinarius elegantior	–	*Cortinarius subtortus*
Cortinarius renidens	–	*Cortinarius callisteus*
Inocybe fastigiata	–	*Inocybe umbrina*
Lactarius ichoratus	–	*Lactarius rufus*
Russula firmula	–	*Russula mustelina*

Krieglsteiner (1977), while studying the macromycetes of mixed silver fir forests of the paenemontaneous area of the Swabian Forest (southwest Germany) stated, that the fungal components of these forests are typical mountain species, especially the lignicolous ones, as known from other areas.

4.2.4. *Fungi in Pinus cembra — Larix decidua forests of central Alps*

Because of the continental climate and the specific qualities of the tree species *Pinus cembra* (five-needle pine) and *Larix decidua*, some associated fungi also are characteristic. Beneath the *Pinus cembra*-associated ectomycorrhizal species, some other macromycetes can be found which are also adapted to other species of the genus *Pinus* (Table VI) or live unspecifically together with coniferous tree species (Table VII) (Singer 1938, Friedrich 1942, Haller 1948, Schärer-Bider 1951, Moser 1956, 1958, Favre 1960, Horak 1963, Trappe 1967, Schmid-Heckel 1985, Tortić 1987). With regard to the litter decomposing and soil-inhabiting fungi, the reported species are unspecific and common species of coniferous forests.

Unspecific ectomycorrhizal species, which can also live together with pine or other coniferous trees are found in *Pinus cembra* forest.

In central Europe, *Larix decidua* is often planted and cultivated in the lowlands, but the distribution area of this species is in continental Eurasia. The fungal flora of artificial larch forest stands is very poor, however, these species can have a high productivity of fruiting bodies. Some specific fungi of *Larix decidua* follow trees to the lowland stands, but in the subalpine area some additional macromycetes enlarge the typical spectrum of *Larix*-associated ectomycorrhizal fungi (Table VIII) (Singer 1938, How 1942, Horak 1963, 1985, Schmid-Heckel 1985, Favre 1955, Dörfelt 1981).

Horak (1963, 1985) designated some other conifer-associated ectomycorrhizal fungi (*Lactarius rufus, Paxillus involutus, Lactarius theiogalus*) which often can be found in these forests. The litter-decomposing and soil-inhabiting fungi are common species, in coniferous forests they are widespread. However, it should especially be noted that the information on the fungal vegetation of the alpine zone is insufficient and should be intensively studied.

4.2.5. *Fungi in pine forest communities*

Pinus sylvestris has a broad ecological amplitude and, therefore, it can grow in a wide range of locations with various climates and soil types. Thus, the

Table VI. Specific ectomycorrhizal species of *Pinus cembra*

Suillus sibiricus
Suillus placidus
Suillus plorans

Table VII. Unspecific ectomycorrhizal species in *Pinus cembra* forest

Xerocomus badius
Lactarius rufus
Chroogomphus helveticus
Lactarius deliciosus
Cortinarius anomalus
Cortinarius salor
Dermocybe semisanguinea
etc.

Table VIII. *Larix*-associated ectomycorrhizal fungi

Widespread larch fungi:	Larch fungi in the subalpine belt:
Boletinus cavipes	*Gomphidius gracilis*
Gomphidius maculatus	*Hygrophorus bresadolae*
Suillus aeruginascens	*Hygrophorus lucorum*
Suillus grevillei	*Hygrophorus speciosus*
Tricholoma psammopus	*Lactarius porninsis*
	Russula laricina
	Suillus tridentinus

pine forest communities are also very dissimilar. Pine forest communities live in dunes on the coast, as well as on submontane to high montane sites in the northern and southern parts of chalky areas of the Alps, they thrive in warm regions together with oak (Querco pubescentis-petreae), as well as at the subarctic timber line in northern Scandinavia together with birch. They can also be found on peaty soil or wet places at the border of moors in the pine forest belts.

Many fungal species which form ectomycorrhiza with pine or saprophytically live in pine forests, are stenecous. In many cases the ecological amplitude of *Pinus sylvestris* is broader than that of the associated fungi. Therefore, the fungal vegetation of certain pine forest is very characteristic.

In Europe, the composition of the macromycetes is very similar in moor pine forests (Table IX) (Neuhoff 1922, Bujakiewicz and Fiklewicz 1963, Einhellinger 1976, 1977, Favre 1948, Kreisel 1954, 1957, Kotlaba and Kubička 1960, Kotlaba 1953, Kotlaba and Pouzar 1962, Nespiak 1959, 1962, Jahn 1954, 1964, Lange 1948, Augustin and Runge 1968, Krisai 1973, Seppälä 1978, Dörfelt 1981, Haas and Kost 1985, Bujakiewicz 1975, 1986, Lisiewska 1978, 1986, Watling 1988). The following macromycetes are characteristic, acidophilic, pine mycorrhizal species: *Suillus flavus, Russula emetica* s. str., *R. helodes, Lactarius helvus, Cortinarius fulvescens, Dermocybe palustris* etc. Additionally, more unspecific conifer-mycorrhized and acid soil-preferring fungi can also be found (*Amanita fulva, Paxillus involutus, Laccaria proxima, Lactarius rufus, L. theiogalus, Scleroderma aurantium*). As is similar in spruce forests, unspecific saprophytic litter-decomposing fungi (*Marasmius androsaceus, Collybia dryophila, C. maculata, Hygrophoropsis aurantiaca, Mycena galopus* etc.) are dominant in pine forests.

The studies of the fungal flora of a Vaccinio-Mugetum in an ombrosoligenous bog (natural reserve: 'Bannwald Waldmoor-Torfstich', northern Black Forest, Germany) revealed ecological differences in which the mycorrhizal partners of pine form fruit-bodies. Some species are adapted to places with moist sphagnum cushions, others choose dry peaty places (Table X). Bryicolous and sphagnicolous fungi of the open bog also appear in wet places in the Vaccinio-Mugetum (*Omphalina* spp., *Galerina* spp., *Hypholoma udum, Lyophyllum palustre*). Inside the Vaccinio-Abietetum, in the neighbourhood of the studied bog, the species composition is significantly dissimilar (Table XI). Einhellinger (1976, 1977) noted 112 species in moist Mugetum of bogs of Bavaria, 79 species of these fungi could also be collected by Favre (1948) in Switzerland in similar habitats. Bujakiewicz and Fiklewicz (1963), Nespiak (1959) and Bujakiewicz (1986) reported an identical arrangement of pine-associated macromycetes from Poland. The data given by Kreisel (1954, 1957), Neuhoff (1922) and Kalamees (1980) support this and prove the conformity of this fungal community (Table IX).

In pine forest on dunes at the coast we can typically find acidic soil preferring, pine associated macromycetes. However, the moist habitats preferring fungi are missing. The species composition of pine forests on sand-dunes described by Massart (1972), Chevassut and Mousain (1973), Bon (1969/70, 1983) show that there are big differences between the forests at the coast and those a few kilometers away from the coast. The pine forest on the dunes are clearly poorer. Studying the fungal vegetations of *Empetrum-Pinus*-forests (Pineto-Empetrum nigri) and the *Pteridium-Pinus*-forests Kreisel (1957) got the same results. In sandy (silicatic) pine forests a rich macro-

Table IX. Characteristic macromycete species of moor pine forests of Europe

	N	F	Ebs	Ep	HK	KK	KP	Ko	K
*Amanita fulva**		F	Eb	Ep	HK	KK			
Cantharellus lutescens				Ep					
Clitocybe vibecina		F		Ep				Ko	
*Collybia dryophila**	N	F	Eb	Ep		KK		Ko	
*Collybia maculata**	N	F		Ep				Ko	
Collybia tuberosa		F		Ep				Ko	
Coprinus stercorarius				Ep					
Cortinarius acutus		F		Ep					
Cortinarius caninus		F		Ep					
Cortinarius cf. *scandens*				Ep					
Cortinarius delibutus		F		Ep	HK				
Cortinarius fasciatus				Ep					
Cortinarius fulvescens		F		Ep	HK				
Cortinarius junghuhnii				Ep					
Cortinarius lucorum				Ep					
Cortinarius muscorum		F							
Cortinarius obtusus		F		Ep					K
*Cortinarius paleaceus**		F	Eb	Ep	HK				
Cortinarius plumbosus		F							
Cortinarius pluvius				Ep					
Cortinarius renidens		F		Ep					
Cortinarius rigidus		F	Es	Ep					
Cortinarius saniosus				Ep					
Cortinarius scaurus		F		Ep	HK				
Cortinarius simulatus				Ep					
Cortinarius speciosissimus		F		Ep	HK				
Cortinarius spilomeus		F		Ep					
Cortinarius strobilaceus				Ep					
Dermocybe cinnamomea					HK	KK			
Dermocybe palustris		F		Ep	HK				
*Dermocybe semisanguinea**		F	Es	Ep	HK	KK			
Dermocybe sphagneti		F		Ep					
Dermocybe sphagnogena		F		Ep					
Dermocybe uliginosa		F		Ep					
Entoloma cetratum	N	F		Ep					
Entoloma staurosporum		F		Ep					
Gomphidius roseus		F		Ep					
Hebeloma longicaudum		F	Eb	Ep					
Hygrocybe cantharellus		F	Es	Ep					
*Hypholoma udum**	N	F		Ep		KK	KP		
Inocybe boltonii		F		Ep					
Inocybe casimiri				Ep					
Inocybe kuehneri				Ep					
Inocybe lanuginella				Ep					
*Inocybe napipes**		F	Eb	Ep	HK			Ko	

Table IX (continued)

	N	F	Ebs	Ep	HK	KK	KP	Ko	K
Inocybe petiginosa		F		Ep					
Inocybe subcaperata		F							
Inocybe terrigena		F		Ep					
*Laccaria amethystea**		F	Es	Ep					
*Laccaria laccata**	N	F	Eb			KK	KP		K
Laccaria proxima		F	Es	Ep					
Lactarius badiosanguineus				Ep					
*Lactarius camphoratus**		F	Eb	Ep	HK				
Lactarius deliciosus		F		Ep					
Lactarius fuliginosus		F		Ep					
*Lactarius helvus**	N	F	Es	Ep	HK	KK			K
Lactarius musteus		F							K
*Lactarius rufus**	N	F		Ep	HK	KK			K
Lactarius sphagneti				Ep					
*Lactarius theiogalus**		F	Eb	Ep	HK			Ko	
Lactarius trivialis		F		Ep					
*Marasmius androsaceus**		F		Ep		KK		Ko	
Micromphale perforans	N	F		Ep					
Mycena adonis		F		Ep					
Mycena alcalina		F		Ep				Ko	
Mycena cinerella		F		Ep					
*Mycena epipterygia**		F	Eb	Ep				Ko	K
Mycena filopes		F		Ep					
*Mycena galopus**		F	Eb	Ep			KP	Ko	K
Mycena phyllogena		F		Ep					
*Mycena rorida**	N	F		Ep				Ko	
*Paxillus involutus**	N	F	Eb	Ep	HK	KK		Ko	K
Pholiota scamba		F		Ep	HK				
*Rickenella fibula**		F	Eb	Ep			KP		
Rozites caperata		F		Ep					
Russula aquosa			Es	Ep					
Russula coerulea				Ep	HK				
*Russula decolorans**	N	F		Ep	HK	KK			K
*Russula emetica**	N	F		Ep	HK	KK			K
Russula fragilis		F		Ep					
Russula helodes					HK				
Russula nitida					HK				
*Russula ochroleuca**		F		Ep	HK			Ko	
Russula olivaceoviolascens			Es	Ep					
*Russula paludosa**	N	F		Ep	HK				K
Russula rhopopoda				Ep					
Russula sardonia				Ep					
Scleroderma citrinum		F				KK		Ko	
*Strobilurus tenacellus**		F		Ep		KK		Ko	
*Suillus bovinus**		F		Ep	HK	KK		Ko	K

Table IX (continued)

	N	F	Ebs	Ep	HK	KK	KP	Ko	K
Suillus flavidus									K
*Suillus variegatus**		F		Ep		KK		Ko	K
Thelephora caryophyllea				Ep					
Thelephora terrestris				Ep	HK				

* = more than four times reported
Eb = Einhellinger (1976, 1977), moist or dry moor forest with *Betula*
Es = Einhellinger (1976, 1977), moist moor forest with *Betula*
Ep = Einhellinger (1976, 1977), Pinetum with *Sphagnum*
F = Favre (1948)
K = Kreisel (1957)
HK = Haas and Kost (1985, 1989)
KK = Kotlaba and Kubička (1960)
KP = Kotlaba and Pouzar (1962)
Ko = Kotlaba (1953)
N = Nespiak (1953)

Table X. Ectomycorrhizal species within Vaccinio-Mugetum (BW-Waldmoor-Torfstich)

In Sphagnum	Sphagnum or peaty soil	Dry peaty soil
mycorrhiza fungi of B E T U L A		
* *Leccinum holopus*	*Cortinarius paleaceus* F	*Amanita fulva*
	Lactarius theiogalus F	*Scleroderma citrinum*
	Russula flava	
	Russula emetica ssp. *betularum*	
mycorrhiza fungi of P I N U S		
Cortinarius. tortuosus	*Lactarius helvus*	*Russula coerulea*
Dermocybe palustris	*Russula decolorans*	*Russula sardonia*
Russula paludosa		*Suillus bovinus*
* *Russula emetica*		*Xerocomus badius* F
* *Russula longipes*		*Lactarius hepaticus*
* *Cortinarius fulvescens*		
other mycorrhiza fungi		
Cortinarius scaurus	*Inocybe napipes*	*Xerocomus parasiticus*
Dermocybe cinnamomea		(*on Sclerod. citrinum*)

F = also associated with *Picea*
* = characteristic for Vaccinio-Mugetum
(Haas and Kost 1985, revised)

Table XI. Ectomycorrhizal species in two neighbouring forest communities of the nature reserve 'Bannwald Waldmoor-Torfstich' (Haas and Kost 1985)

Vaccinio-Mugetum betuletosum 51 species	Vaccinio-Abietetum 24 species
Amanita fulva	***Amanita fulva***
Amanita rubescens	
Boletus edulis	*Cortinarius acutus*
Cortinarius anomalus	*Cortinarius anomalus*
	Cortinarius camphoratus
Cortinarius decipiens	***Cortinarius decipiens***
Cortinarius delibutus	
Cortinarius fulvescens	
Cortinarius gentilis	*Cortinarius obtusus*
Cortinarius paleaceus	***Cortinarius paleaceus***
Cortinarius pholideus	
Cortinarius scaurus	***Cortinarius scaurus***
Cortinarius speciosissimus	***Cortinarius speciosissimus***
Cortinarius tortuosus	
Cortinarius turibulosus	
Dermocybe cinnamomea	***Dermocybe cinnamomea***
Dermocybe malicoria	
Dermocybe palustris	
Dermocybe semisanguinea	*Inocybe acuta*
Hygrophorus olivaceoalbus	*Inocybe lanuginosa*
Inocybe napipes	***Inocybe napipes***
Inocybe umbrina	*Laccaria amethystea*
	Laccaria proxima
Lactarius camphoratus	***Lactarius camphoratus***
Lactarius helvus	
Lactarius hepaticus	
Lactarius lignyotus	
Lactarius necator	***Lactarius necator***
Lactarius rufus	
Lactarius theiogalus	***Lactarius theiogalus***
Leccinum holopus	
Leccinum oxydabile	***Leccinum oxydabile***
Leccinum scabr. coloratipes	
Leccinum scabrum	
Leccinum variecolor	
Paxillus involutus	***Paxillus involutus***
Russula coerulea	
Russula decolorans	
Russula emetica	
Russula emetica s. *betularum*	
Russula emetica s. *longipes*	
Russula flava	
Russula helodes	
Russula nitida	***Russula nitida***
Russula ochroleuca	***Russula ochroleuca***

Table XI (continued)

Vaccinio-Mugetum betuletosum 51 species	Vaccinio-Abietetum 24 species
Russula paludosa	
Scleroderma citrinum	***Schleroderma citrinum***
Suillus bovinus	
Thelephora terrestris	
Xerocomus badius	
Xerocomus parasiticus	
Xerocomus spadiceus	

mycete flora thrives (Nathorst-Windahl 1956, Nespiak 1959, Haas 1963, Ulvinen 1963, Guminska 1976, 1989, Ulvinen *et al.* 1981, Dörfelt 1981, Kreisel 1981, Moser 1982, Ferrari *et al.* 1982). Here, important genera are *Suillus* spp., *Tricholoma* spp., the stipitate, hydnaceous fungi (*Sarcodon* spp., *Hydnellum* spp., *Bankera* spp., *Phellodon* spp.) and many others. Arnolds (1985, 1988a,b, 1989a,b,c) pointed out that many macrofungi of pine forests on poor, acid soil (*Cladonia*-pine forest) are strongly threatened. The decline of hydnaceous and other mycorrhizal fungi is mainly ascribed to nitrogen accumulation in forest soils coming from air pollution (Arnolds 1989c).

In pine plantations the number of fungal species is strongly reduced and only the most common macromycetes appear at these stands (Wilkins and Harris 1946, Bourgeois 1952, Dennis 1955, Heinemann and Darimont 1956, Kreisel 1957, Géhu 1960, Tüxen and Jahns 1964, Šmarda 1965, Richardson 1970, Courtois 1979). Afforested pine forests on locations of natural oak woods have a macromycete flora which contains typical elements of the oak woods (Šmarda 1965, Kost and Haas 1989). The effects of forest management procedures on fruit-body production and macromycete composition in pine forests are comparable with spruce stands (see above) (Ritter and Tölle 1978, Menge and Grand 1978).

The *Pinus sylvestris* forests on chalky soil have a specific and very characteristic fungal composition (Neuhoff 1956, Jahn 1957/58, Höfler 1962, Winterhoff 1977, Bon 1969/70, 1983, Dörfelt 1981). Some pine-associated mycorrhizal species (*Cortinarius russeoides, Hebeloma edurum, Hygrophorus fuscoalbus, Russula sanguinea, Tricholoma batschii, T. terreum, Suillus collinitus, Suillus granulatus*, etc.) can only be found on chalky soil. Winterhoff (1977) studied macromycetes of Pyrolo-Pinetum on chalky sand dunes in the river Rhine plain and reported the characteristic fungal composition on these stands. He stated, that some acidiphilic agarics are able to fructificate on that small places, on which the chalk is washed out and the soil is acidic. The data of macromycetes of alpine Erico-Pinetum silvestris — and Pyrolo-Pinetum — forests of Switzerland are similar to other pine forests (Horak 1985). Hintikka (1988) investigated the macromycete flora in oligotrophic pine forests of different ages in south Finland.

Some information of macromycetes of *Pinus nigra* forests in Austria are given by Friedrich (1940), Höfler and Cernohorsky (1954) and Sprongl (1951). From *Pinus peuce*-forests in Bulgaria first reports are published by Gyosheva and Bogoev (1988) and Khinkova and Drumeva (1978).

In Florida (North-America) Murrill (1949) investigated the various fungal vegetation of high pine woods (*Pinus palustris, Pinus taeda*). Some data of macromycete composition of pine plantations in tropical areas are published by Singer and Moser (1964), Ivory (1980) and Kost (1991).

References

Agerer, R., and I. Kottke, 1981. Sozio-ökologische Studien in Fichten- und Eichen-Buchen-Hainbuchen-Wäldern im Naturpark Schönbuch. Z. Mykol. 47: 103—122.

Al Abras, K., I. Bilger, F. Martin, F. Le Tacon, and F. Lapeyrie. 1988. Morphological and physiological changes in ectomycorrhizas of spruce. New Phytol. 110: 535—540.

Alexander, I. J. 1981. The *Picea sitchensis* + *Lactarius rufus* mycorrhizal association and its effects on seedling growth and development. Trans. Br. mycol. Soc. 76: 417—423.

Alexander, J., and R. Watling. 1987. Macrofungi of Sitka spruce in Scotland. Proc. Roy. Soc. Edinburgh 93B: 107—115.

Arnolds, E. 1985. Veranderingen in de paddestoelenflora (Mykoflora). Wetensch. meded. Kon. ned. natuurhist. ver. 167.

Arnolds, E. 1988a. The changing macromycete flora in the Netherlands. Trans. Brit. mycol. Soc. 90: 391—406.

Arnolds, E. 1988b. The Netherlands as an environment for agarics and boleti. p. 6—28. *In*: C. Bas *et al.* Flora agaricina neerlandica I.

Arnolds, E. 1989a. A preliminary red data list of macrofungi in the Netherlands. Persoonia 14: 77—125.

Arnolds, E., and B. de Vries 1989b. Oecologische statistiek van de Nederlandse macrofungi. Coolia 32: 76—86.

Arnolds, E. 1989c. Former and present distribution of stipitate hydnaceous fungi (Basidiomycetes) in the Netherlands. Nova Hedwigia 48: 107—142.

Augustin, A., and A. Runge. 1968. Pilze in Scheiden-Wollgras-Rasen des Emsdettener Venns. Natur und Heimat 28: 152—153.

Baxter, D. V. 1937. Development and succession of forest fungi and diseases in forest plantations. Univ. Mich. School of Forestry and Conservation Circ. 1: 1—45.

Bon, M. 1969/1970. Flore héliophile des macromycètes de la zone maritime picarde. Thèse. Université Lille.

Bon, M. 1983. Les principaux biotopes du nord de la France. Association multidisciplinaire des biologistes de l'environnement A.M.B.E. Actes du colloque. Le patrimonine naturel régional Nord-Pas-de-Calais: 117—125.

Bon, M., and J.-M. Gehu. 1973. Unités supérieures de végétation et récoltes mycologiques. Doc. Mycol. 2: 1—40.

Boudier, E. 1901. Influence de la nature du sol et des végétaux qui y croissent sur le développement des champignons. Bull. Soc. Mycol. France 17: 55—71.

Bourgeois, O. 1952. Influence des plantations de conifères sur la flore mycologique de la région de Dijon. Publ. Univ. Dijon nouvelle série 9: 27—51.

Bujakiewicz, A. 1975. Grzyby wyszse lasów pszczyńskich. Badania Fizjograficzne nad Polska Zachodina 28 B. Botanika: 25—47.

Bujakiewicz, A. 1982. Macromycetes as an element of forest structure on the Babia Góra massif. p. 645—656. *In*: H. Dierschke (ed.) Struktur und Dynamik von Wäldern. Berichte der Internationalen Symposien der Internationalen Vereinigung für Vegetationskunde, Rinteln (13.—16. 4. 1982).

Bujakiewicz, A. 1986. Udział Macromycetes w zbiorowiskach roślinnych występujacych na podłożu torowym w słowińskim parku narodowym. Badania Fizjograficzne nad Polsak Zachodina 37 B. Botanika: 101—129.

Bujakiewicz, A., and G. Fiklewicz. 1964. Obserwacje fenologczno-ekologiczne nad gzrybami wyzszymi w gradach okolic opalenicy (Zachodnia Wielkopolska). Poznanskie Towarzystwo Przyjaciól Nauk. Wydzial matematyczno-Przyrodniczny, Prace Komisji Biologicznej 26: 13—69.

Carbiener, R., N. Ourisson, and A. Bernard. 1975a. Erfahrungen über Beziehungen zwischen Großpilzen und Pflanzengesellschaften in der Rheinebene und den Vogesen. Beitr. naturk. Forsch. Südw.-Dtl. 34 (Oberdorferfestschrift): 37—56.

Carbiener, R., N. Ourisson, and A. Bernard. 1975b. Premières notes sur les relation entre la répartition des champignons supérieurs et celle des groupements végétaux dans les forêts de la plaine d'Alsace entre Strasbourg et Sélestat. Bull. Soc. Hist. Nat. Colmar 55: 3—36.

Chevassut, G., and D. Mousain. 1973. La macroflore fongique du pin maritime: Essai d'analyse mycosociologique de deux stations de pin maritime dans la région de Montpellier. Bull. Soc. Myc. France 89: 229—251.

Courtois, H. 1979. Two different fungal groups in 55-year-old Scotch pines on a groundwater-influenced sandy site. Angew. Bot. 53: 255—260.

Darimont, F. 1973. Recherches mycosociologiques dans les forêts de Haute Belgique. Mem. Inst. Royal Sci Nat. Belgique 170 (I): 1—220, (II): 68 tables.

Dennis, R. G. W. 1955. The larger fungi in the north-west highlands of Scotland. Kew Bull. 10: 111—126.

Dighton, J., and P. A. Mason. 1985. Mycorrhizal dynamics during forest tree development. p. 117—139. *In*: D. Moore *et al.* (eds.) Developmental biology of higher fungi. British Mycological Society Symposium 10. Cambridge University Press.

Dighton, J., J. M. Poskitt, and D. M. Howard. 1986. Changes in occurrence of basidiomycete fruit bodies during forest stand development with specific reference to mycorrhizal species. Trans. Br. mycol. Soc. 87: 163—171.

Dörfelt, H. 1974a. Die Erforschung der Mycocoenosen als Elemente der Ökosysteme. Mitt. Sekt. Geobotanik und Phytotaxonomie der Biolog. Ges. der DDR. Sonderheft 'Grundlagen der Ökosystemforschung': 85—91.

Dörfelt, H. 1974b. Zur Frage der Beziehungen zwischen Mykocoenosen und Phytocoenosen. Archiv Naturschutz und Landschaftsforsch. 14: 225—228.

Dörfelt, H. 1981. Pilzsoziologie. p. 77—97. *In*: E. Michael, B. Hennig (begründet), and H. Kreisel (eds.) Handbuch für Pilzfreunde IV. Jena.

Drumeva, M., and T. Khinkova. 1978a. The Macromycetes in some pine plantations of Bulgaria. Fitologija (Sofia) 10: 71—85.

Eddelbüttel, H. 1911. Grundlagen einer Pilzflora des östlichen Weserberglandes und ihrer pflanzengeographischen Beziehungen. Ann. Mycol. 9: 445—529.

Einhellinger, A. 1976. Die Pilze in primären und sekundären Pflanzengesellschaften oberbayerischer Moore. Teil 1. Ber. Bayer. Bot. Ges. 47: 75—149.

Einhellinger, A. 1977. Die Pilze in primären und sekundären Pflanzengesellschaften oberbayerischer Moore. Teil 2. Ber. Bayer. Bot. Ges. 48: 61—146.

Engel, H., and I. Friederichsen. 1971. Der Aspekt der Großpilze im Nadelwaldgürtel der nördlichen Kalkalpen in Tirol. I. Die Artenliste und ihre Veränderungen. Z. Pilzk. 37: 1—73.

Engel, H., and I. Friederichsen. 1974. Der Aspekt der Großpilze im Nadelwaldgürtel der nördlichen Kalkalpen in Tirol. II. Die Artenliste. Z. Pilzk. 40: 25—68.

Engel, H., and I. Friederichsen. 1976. Der Aspekt der Großpilze um Mitte September im Nadelwaldgürtel der nördlichen Kalkalpen in Tirol. III. Die Arten am Piller in den westlichen Öztaler Alpen und ein Vergleich der einzelnen Exkursionsgebiete. Z. Pilzk. 42: 79—94.

Engel, H., G. J. Krieglsteiner, A. Dermek, and R. Watling. 1983. Dickröhrlinge. Die Gattung *Boletus* in Europa. Weidhausen.

Favre, J. 1948. Les associations fongiques des hauts-marais jurassiens. Matériaux Flore Crypt. Suisse 10: 3.

Favre, J. 1955. Les champignons supérieurs de la zone alpine du Parc National Suisse. Liestal.

Favre, J. 1960. Catalogue descriptif des champignons supérieurs de la zone subalpine du Parc National Suisse. Ergebnisse wiss. Unters. Schweiz. Nationalparks 6: 321—610.

Fellner, R. 1987. Poznámky k mykocenologické syntaxonomii. 1. Zásady výstavby syntaxonomické klasifikace mykocenóz. Česká Mykologie 41: 225—231.

Fellner, R. 1988. Poznámky k mykocenologické syntaxonomii. 2. Přeheled syntaxonomické klasifikace mykocenóz respektujici zásadu jednoty substrátu a trofismu. Česká Mykologie 42: 41—51.

Ferrari E., A. Venturini, and A. Zuccherelli. 1982. I funghi delle pinete di Ravenna. Ravenna.

Ferry, R. 1887. Espèces acicoles et espèces foliicoles. Rev. Mycol. 9: 42—47.

Ferry, R. 1892. Espèces calcicoles et espèces silicicoles. Rev. Mycol. 14: 146—155.

Fiedler, H.-J., and W. Hunger. 1963. Über den Einfluß einer Kalkdüngung auf das Vorkommen, Wachstum und Nährelementgehalt höherer Pilze im Fichtenbestand. Arch. Forstw. 12: 936—962.

Friedrich, K. 1940. Untersuchungen zur Ökologie der höheren Pilze. Pflanzenforschung 22: 53 pp.

Friedrich, K. 1942. Pilzökologische Untersuchungen in den Ötztaler Alpen. Ber. Deutsch. Bot. Ges. 60: 218—231.

Fries, E. M. 1921—1832. Systema mycologicum, sistens fungorum ordines, genera, huiusque cognitae, quas ad normam methodi naturalis determinavit. Disposuit atque descripsit. Vol. I, 1981, Vol. II. 1822—1823, Vol. III, 1829—1832.

Gilbert, E.-J. 1928. La mycologie sur la terrain. Librairie le Françoise, Paris.

Gilli, A. 1951. Basidiomyzeten der Nadelwälder in einem Gebiet ohne spontane Nadelbäume. Sydowia 5: 129—134.

Graham, V. O. 1927. Ecology of the fungi of the Chicago region. Bot. Gaz. 83: 267—287.

Guminska, B. 1976. Macromycetes łąk w Pienińskim Parku Narodowym. Acta Mycol. 12: 3—75.

Guminska, B. 1989. Macromycetes of the Pieniny national park. (a guide). Abstr. 19th Intern. Phytogeographic Exkursion 1989 Flora and Vegetation of Poland — changes, management and conservation: 1928—1988: 1—34.

Gyosheva, M. M., and V. M. Bogoev. 1988. Mycological investigation into two Balkan pine stands of the Vitosha National Park. Annuaire de l'universite de Sofia 'Kliment Ohridski', Faculte de Biologie 2, 1985 (publ. 1988): 64—78.

Haas, H. 1932. Die bodenbewohnenden Großpilze in den Waldformationen einiger Gebiete von Württemberg. Beih. Bot. Centralbl. 50 Abt. II: 35—134.

Haas, H. 1953. Pilzkunde und Pflanzensoziologie. Z. Pilzk. 13: 1—5.

Haas, H. 1958. Die Pilzflora der Tannenmischwälder an der Muschelkalk-Buntsandstein-Grenze des Ost-Schwarzwaldes. Z. Pilzk. 24: 61—67.

Haas, H. 1963. Dritter europäischer Mykologenkongress in Glasgow vom. 1.—7. September 1963. Z. Pilzk. 29: 45—47.

Haas, H. 1971. Makromyzetenflora und Kernfäulebefall älterer Fichtenbestände auf der

Schwäbischen Alb. Mitt. Ver. Forstl. Standortskde. u. Forstpflanzenzüchtung 20: 50—59.
Haas, H. 1972. Beiträge zur Kenntnis der Pilzflora im Raum zwischen Brigach, Eschach und Prim. Schriften Ver. Geschichte u. Naturgesch. d. Baar 29: 145—201.
Haas, H. 1975. Die Pilzflora in rotfäulegefährdeten Fichtenbeständen der Schwäbischen Alb. Z. Pilzk. 41: 45—54.
Haas, H. 1979. Die Pilzflora in rotfäulebefallenen Fichten-Durchforstungsbeständen der Schwäbischen Alb. Mitt. Ver. Forstl. Standortskde. u. Forstpflanzenzüchtung 27: 6—25.
Haas, H., and G. Kost. 1985. Basidiomycetenflora des Bannwaldes Waldmoor-Torfstich. 'Waldschutzgebiete' im Rahmen der Mitteilungen der forstlichen Versuchs- und Forschungsanstalt. 3. Der Bannwald 'Waldmoor-Torfstich' 105—124.
Haller, R. 1948. Quelques observations sur les bolets des forêts d'Epiceas et de Mélèzes. Schweiz. Z. Pilzk. 26: 77—79.
Heinemann, P., and F. Darimont. 1956. Premières indications sur les relations entre les Champignons et les groupements végéteaux de Belgique. Les Naturalistes Belges. (Publ. prem. Session Européenne de Mycologie): 25—39.
Hintikka, V. 1988. On the macromycete flora in oligotrophic pine forest of different ages in South Finland. Acta Bot. Fennica 136: 89—94.
Höfler, K. 1938. Pilzsoziologie. Ber. Deutsch. Bot Ges. 55: 606—622.
Höfler, K. 1955a. Zur Pilzvegetation aufgeforsteter Fichtenwälder. Sydowia 9: 246—255.
Höfler, K. 1955b. Über Pilzsoziologie. Verhandl. Zool.-Bot. Ges. Wien 95: 58—75.
Höfler, K. 1956. Über Pilzsoziologie. Z. Pilzk 22: 42—54.
Höfler, K. 1962. Sommerliche Pilzaspekte um Bayreuth. Z. Pilzk. 28: 1—5.
Höfler, K., and T. Cernohorsky. 1954. Pilzexkursion auf den Mödlinger Frauenstein. Verh. Zool.-Bot. Ges. Wien 94: 159—164.
Hora, F. B. 1959. Quantitative experiments on toadstools production in woods. Trans. Brit. mycol. Soc. 42: 1—14.
Hora, F. B. 1972. Productivity of toadstools in coniferous plantations natural and experimental. Mycopath. Mycol. Appl. 48: 35—42.
Horak, E. 1963. Pilzsoziologische Untersuchungen in der subalpinen Stufe (Piceetum subalpinum und Rhodoreto-Vaccinietum) der Rhätischen Alpen. Mitt. Schweiz. Anst. Forstl. Vers.-Wes. 39: 1—112.
Horak, E. 1985. Ökologische Untersuchungen im Unterengadin. Ergeb. wiss. Unters. Schweiz. Nationalpark 12: 337—476.
Horak, E., and O. Röllin. 1989. Der Einfluß von Klärschlamm auf die Mykorrhizaflora eines Eichen-Hainbuchenwaldes bei Genf. Mitt. eidgen. Anst. forstl. Vers.-w. 64: 21—147.
How, J. E. 1942. The mycorrhizal relation to larch: III Ann. Bot. N.S. 6: 103—126.
Ivory, M. H. 1980. Ecotmycorrhizal fungi of lowland tropical pines in natural forests and exotic plantations. p. 110—117. *In*: P. Mikola (ed.) Tropical Mycorrhiza research. Oxford.
Jahn, H. 1954. Zur Pilzflora des Naturschutzgebietes 'Heiliges Meer'. Natur und Heimat 14: 97—115.
Jahn, H. 1957/58. Die Täublinge (*Russula*) der nordwestdeutschen Kiefernforsten im westfälischen Raum. Westf. Pilzbr. 1: 6—10.
Jahn, H. 1964. Das Sumpfgraublatt, *Lyophyllum palustre* (Peck)-Singer. Westf. Pilzb. 5: 13—15.
Jahn, H. 1968. Pilze an Weißtanne. Westf. Pilzb. 7: 17—40.
Jahn, H. 1969. Zur Pilzflora der subalpinen Fichtenwälder (Piceetum subalpinum) im Oberen Harz. Westf. Pilzb. 7: 93—102.
Kalamees, K. 1980. Ecology and distribution of Fungi (Agaricales, Helotiales, Erysiphales, Gasteromycetes). Scripta Mycologica 9: 1—98.
Khinkova, T., and M. Drmeva. 1978b. Materials concerning Ascomycetes species composition and distribution in Bulgaria. II. Fitologija 10: 71—85.
Kost, G. 1991a. Veränderungen im Artenspektrum der Pilze in Vergangenheit und Zukunft. Veröff. Nordd. Naturschutzakademie (In press).

Kost, G. 1991b. Contribution to the fungal flora of East Africa I. Introduction and survey. Nova Hedwigia (In press).
Kost, G., and H. Haas. 1989. Die Pilzflora von Bannwäldern in Baden-Württemberg. Ein Beitrag zur Kenntnis der Vergesellschaftung höherer Pilze in einigen Waldgesellschaften Süddeutschlands. 'Waldschutzgebiete' im Rahmen der Mitteilungen der forstlichen Versuchs- und Forschungsanstalt 4: 9—182.
Kotlaba, F. 1953. Ekologicko-sociologická studie o mycoflore 'Sobeslvskych blat'. Preslia 25: 305—350.
Kotlaba, F., J. Kubicka 1960. Die Mykoflora des Moores 'Rotes Moos' bei Schalmanowitz in ihrer Beziehung zur Mykoflora der südböhmischen Torfgebiete. Česká Mykol. 14: 90—100.
Kotlaba, F., and Z. Pouzar. 1962. Lupenté a hřibovité houby (Agaricales) Dobročského pralesa na Slovensku. Česká Mykologie 17: 71—76.
Kreisel, H. 1954. Beobachtungen über die Pilzflora einiger Hoch- und Zwischenmoore Ost-Mecklenburgs. Wiss. Z. Univ. Greifswald, Math.-Nat. Reihe 3: 291—300.
Kreisel, H. 1957. Die Pilzflora des Darß und ihre Stellung in der Gesamtvegetation. Feddes Repert. Beih. 137: 110—183.
Krieglsteiner, G. J. 1977. Die Makromyzeten der Tannen-Mischwälder des Inneren Schwäbisch-Fränkischen Waldes (Ostwürrtemberg). Lempp, Schwäbisch Gmünd.
Krisai, D. 1973. Höhere Pilze aus dem Trummer Seen-Gebiet. Mitt. Bot. Linz 5: 206—214.
Kuyper, T. 1988. The effects of forest fertilization on abundance and diversity of ectomycorrhizal fungi. p. 146—149. *In*: A. E. Jansen, *et al.* (eds.) Ectomycorrhiza and acid rain, Commission of the European Communities, Brussels.
Kuyper, T. 1989. Auswirkung der Walddüngung auf die Mykoflora. Beitr. zur Kenntnis der Pilze Mitteleuropas 5: 5—20.
Laber, P., and D. Laber. 1988. Die Pilzflora des Belchengebietes. Natur- und Landschaftsschutzgebiete in Baden-Württemberg 13, Der Belchen — Geschichtlich-naturkundliche Monographie des schönsten Schwarzwaldberges: 555—592.
Lange, M. 1948. The agarics of Maglemoose. Dansk Bot. Ark. 13: 1—141.
Lisiewska, M. 1978. Maromycetes na tle zespolów lesnych Śietokrzyskiego parku Narodowego. Acta Mycol. 14: 163—191.
Lisiewska, M. 1986. Gryzby wyzsze Wolinsko parku Narodowego. Higher fungi of the Wolin island national park. Acta Mycologica 2: 25—77.
Malençon, G. 1951. Climat et mycologie. Bull. Soc. Nat. d'Oyonnax 5: 53—71.
Marschner, H., V. Römheld, W. J. Horst, and P. Martin. 1986. Root-induced changes in the rhizosphere: Importance for the mineral nutrition of plants. Z. Pflanzenernaehr. Bodenk. 149: 441—456.
Massart, F. 1972. Aspect de la flore fongique des dunes à couvert de pins maritimes du Porge. Bull. Soc. Linn. Bordeaux 2(4).
Meisel-Jahn, S., and W. Pirk. 1955. Über das soziologische Verhalten von Pilzen in Fichten-Forstgesellschaften. Mitt. Florist.-soz. Arbeitsgem. 5: 59—63.
Menge, J. A., and L. F. Grand. 1978. Effect of fertilization on production of epigeous basiodiocarps of mycorrhizal fungi in loblolly pine plantations. Can. J. Bot. 56: 2357—2362.
Moser, M. 1956. Die Bedeutung der Mykorrhiza für Aufforstungen in Hochlagen. Forstwiss. Centralbl. 75: 8—18.
Moser, M. 1958. Die künstliche Mykorrhizaimpfung an Forstpflanzen. I. Erfahrung bei der Reinkultur von Mykorrhizapilzen. Forstwiss. Centralbl. 77: 32—40.
Moser, M. 1959. Pilz und Baum. Z. Pilzk. 27: 37—53.
Moser, M. 1982. Mycoflora of the transitional zone from subalpine forests to alpine tundra. p. 371—389. *In*: G. A. Laursen, and J. F. Ammirati (eds.) Arctic and Alpine Mycology. The first international Symposium on arcto-alpine mycology. University of Washington Press, Seattle and London.

Murrill, W. A. 1949. Terrestrial basidiomycete fungi of the Florida high pine woods. Ecology 30: 377—382.
Nathorst-Windahl, T. 1956. Zur Verbreitung der Agaricales in den Wäldern des südwestlichen Schwedens. Friesia 5 (1954—1956): 319—324.
Nespiak, A. 1959. Studia nad udzialem grzybów kapeluszowych w zespolach lesnych na terenie bialowieskiego parku narodowego. Monographiae Botanicae 8: 3—141.
Nespiak, A. 1959. The investigations on the character of the correlations between the higher fungi and wood associations in the National Park Bialowieza. Monographiae Botanicae 8: 3—141.
Nespiak, A. 1960. Notatki mikologiczne z Tatr. Notes mycologiques de Tatra. Fragm. Flor. et Geobot. 6: 709—724.
Nespiak, A. 1962. Observations sur les champignons à chapeau dans les associations forestières en Pologne. Vegetatio 11: 71—74.
Nespiak, A. 1971. Grzyby wyżsne regla górnego w Karkonoszach. Die Pilze in dem Piceetum herecynicum in Karkonosze. Acta Mycol. 7: 87—98.
Neuhoff, W. 1922. Wanderungen zum Zehlau-Hochmoor. Z. Pilzk. 1: 54—58.
Neuhoff, W. 1956. Die Milchlinge (*Lactarii*). Die Pilze Mitteleuropas. Vol. 2b. Bad Heilbrunn.
Oberdorfer, E. 1987. Süddeutsche Wald- und Gebüschgesellschaften im europäischen Rahmen. Tuexenia 7: 459—468.
Ohenoja, E. 1978. Kuusamon siemitutkimuksesta. Acta Univ. Ouluensis A 68 Biol 4: 97—105.
Ohenoja, E. 1983. Lannoituksen vaikutuksesta kangasmetsien syyssienisatoon Pudasjärvellä vuosina 1979—1980. Manuskript, University of Oulu.
Ohenoja, E. 1988a. Effect of forest management procedures on fungal fruitbody production in Finland. Acta Bot. Fennica 136: 81—84.
Ohenoja, E. 1988b. Behaviour of mycorrhizal fungi in fertilized forests. Karstenia 28: 27—30.
Persoon, C. H. 1822—1828. Mycologia europaea s. completa omnium fungorum in variis Europae regionibus detectorum enumeratio, methodo naturali disposita, descriptione succincta, synonyma selecta et observationibus criticis additis elaborata. Sect. I, 12pl. col., Erlangen, 1822, Sect. II, 214 pp., 10pl., 1825, Sect. III, 282 pp., 9pl., 1928.
Pirk, W. 1944. Zur Soziologie der Pilze im Querceto-Carpinetum. Rundbrief 14: 1—10.
Pirk, W. 1948. Zur Soziologie der Pilze im Querceto-Carpinetum. Z. Pilzk. 21: 11—20.
Ricek, E. W. 1961. Beiträge zu einer Pilzflora des Attergaues in Österreich. Sydowia 15: 159—184.
Ricek, E. W. 1981. Die Pilzgesellschaften heranwachsender Fichtenbestände auf ehemaligen Wiesenflächen. Z. Mykol. 47: 123—148.
Richardson, M. J. 1970. Studies on *Russula emetica* and other agarics in a scots pine plantation. Trans. Brit. mycol. Soc. 55: 217—229.
Ritter, G., and H. Tölle. 1978. Stickstoffdüngung in Kiefernbeständen und ihre Wirkung auf Mykorrhizabildung und Fruktifikation der Symbiosepilze. Beitr. für die Forstwirtschaft 12: 162—166.
Rücker, T., and T. Peer. 1988. Pilzsoziologische Untersuchungen am Stubnerkogel (Gasteiner Tal, Salzburg, Österreich) unter Berücksichtigung der Schwermetallsituation. Nova Hedwigia 47: 1—38.
Salo, K. 1979. Mushrooms and mushroom yield on transitional peatlands in central Finland. Ann. Bot. Fenn. 16: 181—192.
Schärer-Bider, W. 1951. Beobachtungen über die Verbreitung einiger höherer Pilze im Wallis. Ber. Geobot. Forsch. Inst. Rübel 1950: 38—44.
Scharfetter. 1908. Eine Pilzausstellung am Stadtgymnasium in Villach, nebst Anmerkungen zur Ökologie der Höheren Pilze. Carinthia (Mitt. naturhist. Mus. Kärnten) 2: 106—124.
Schmid-Heckel, H. 1985. Zur Kenntnis der Pilze in den Nördlichen Kalkalpen. Nationalpark Berchtesgaden. Forschungsber. 8. Herausg. Nationalparkverwaltung Berchtesgaden.
Schwöbel, H. 1971. Beitrag zur Kenntnis der Pilzflora des Wutachgebietes. p. 227—238. *In*: K. F. J. Sauer, and M. Schnitter (eds.) Die Wutach. Die Natur- und Landschaftschafts-

schutzgebiete Baden-Württembergs 6. Badischer Landesverein für Naturkunde und Naturschutz e.V., Freiburg.

Seibt, G. 1981. Die Buchen- und Fichtenbestände der Probeflächen des Solling Projektes der Deutschen Forschungsgemeinschaft. Schriften der Forstl. Fak. Univ. Göttingen 72: 1—109.

Seidel, D. 1983. Vegetionsanalytische Daten als Grundlage für die Beurteilung von Interdependenzen zwischen Moosen und Pilzen. Beitr. Biol. Pflanzen 58: 95—114.

Seppälä, K. 1978. Effect of dwarf-scrub vegetation suppression on berry and mushroom yield on a drained pine swamp. Suo 29: 69—74.

Singer, R. 1938. Über Lärchen-, Zirben- und Birkenröhrlinge. Schweiz. Zeitschr. Pilzk. 16: 123—126.

Singer, R., and M. Moser. 1964. Forest mycology and forest communities in South America. Mycopath. et Mycol. Appl. 36: 129—191.

Šmarda, F. 1965. Mykocenologické srovnání borů na pųesypoých píscích Dolonomoravského úvalu na jižní Moravě a v Záhorské nížině na západní Slovensku. Česka Mykologie 19: 11—20.

Šmarda, F. 1972. Pilzgesellschaften einiger Laubwälder Mährens. Acta Sc. Nat. Brno 6: 1—53.

Šmarda, F. 1973. Die Pilzgesellschaften einiger Fichtenwälder Mährens. Act. Natural. Acad. Scient. Bohemosl. Brno Nova series 7: 1—44.

Sprongl, K. 1951. Beiträge zur Pilzflora des Gaadener Becken in Niederösterreich. Sydowia 5: 135.

Stricker, P. 1950. Der Pilzbestand der Wutachschlucht, einiger Seitenschluchten und der angrenzenden Wälder. Beitr. Naturk. Forschung Südwest-Dtl. 9: 3—54.

Svrček, M. 1960. Eine mykofloristische Skizze der Umgebung von Karlštejn (Karlstein) im Mittelböhmen. Česka Mykol. 14: 67—86.

Thomas, G. W., D. Rogers, and Jackson. 1983. Changes in the mycorrhizal status of Sitka spruce following outplanting. Plant and Soil 71: 319—323.

Tortić, M. 1986. Betrachtungen über Mykorrhiza-Pilze fünfnadeliger Kiefern. Beiträge zur Kenntnis der Pilze Mitteleuropas 3: 71—78.

Trappe, J. M. 1967. Fungus associates of ectotrophic mycorrhiza. Bot. Rev. 28: 538—606.

Tüxen, R., and W. Jahns 1964. Kurzer Bericht über die Ergebnisse der pilzsoziologischen Tagung am 28./29. Oktober 1963 in Stolzenau/Weser. (Als Manuskript vervielfältigt, Stolzenau/Weser).

Uebel, E. 1982. Einfluß einer Mineraldüngung auf höhere Pilze und die Mykorrhizabildung auf einer aufgeforsteten Ackerfläche. Arch. Naturschutz u. Landschaftsforsch. 22: 169—175.

Ulvinen, T. 1963. Über die Großpilze der Oulu-Gegend. Aquilo 1: 38—52.

Ulvinen, T., E. Ohenoja, and T. Ahti. 1981. A checklist of the fungi (incl. lichens) of the Koillisma Kuusamo biological province N.E. Finland. Oulanka Reports 2: 1—64.

Walter, H. 1951. Einführung in die Phytologie. III. Grundlagen der Pflanzenverbreitung, 1. Teil: Standortslehre (analytisch-ökologische Geobotanik). Ulmer, Stuttgart.

Watling, R. 1988. A mycological kaleidoscope. Trans. Br. mycol. Soc. 90: 1—28.

Wilkins, W. H., and G. C. M. Harris. 1946. The ecology of the larger fungi V. An investigation into the influence of rainfall and temperature on the seasonal production of fungi in a beechwood and a pinewood. Ann. Appl. Biol. 33: 179—188.

Winterhoff, W. 1977. Die Pilzflora des Naturschutzgebietes Sandhausener Dünen bei Heidelberg. Veröff. Landesstelle Naturschutz und Landschaftspflege Baden-Württ. 44/45: 51—118.

Winterhoff, W. 1984a. Analyse der Pilze in Pflanzengesellschaften, insbesondere der Makromyceten. p. 227—248. *In*: R. Knapp (ed.) Sampling Methods and taxon analysis in vegetation science. Handbook of vegetation science Vol. 4. Dr. W. Junck Publishers, The Hague.

Winterhoff, W., and G. J. Krieglsteiner. 1984b. Gefährdete Pilze in Baden-Württemberg. Beih. Veröff. Landesstelle Naturschutz und Landschaftspflege Baden-Württ. 40: 1—116.

Winterhoff, W. 1984c. Vorläufige Rote Liste der Großpilze (Makromyzeten). p. 162—184. *In*: J. Blab, E. Nowak, W. Trautmann, and H. Sukopp (eds.). Rote Liste der gefährdeten Tiere und Pflanzen in der Bundesrepublik. Naturschutz Aktuell Nr. 1. Kilda, Greven.

5. Macrofungal communities outside forests

EEF ARNOLDS

Summary

Data are summarized on macrofungocoenoses in plant communities dominated by herbs or dwarf shrubs, in particular grasslands, heathlands, bogs, alpine communities and ephemeral weed communities. Some special methodological problems in these habitats are discussed. The functions of macrofungi as saprophytes, parasites and mutualistic symbionts on different substrates are reviewed. Attention is paid to fungi associated with bryophytes. Data are provided on the species diversity and abundance and productivity of sporocarps in different plant communities. The relations between macrofungocoenoses and environmental conditions (soil, climate, and human influence) are discussed. Finally, short characteristics are given of the macrofungocoenoses in various plant communities with references to the most relevant literature.

5.1. Introduction

The aim of this contribution is to summarize the present knowledge on macrofungocoenoses in plant communities, dominated by herbs or dwarf-shrubs up to about 0.5 metres tall, comprising aquatic communities, communities of weeds and ruderals, pioneer communities, grasslands, bogs, marshes, shores, heathlands and alpine communities.

Mushrooms and toadstools are often intuitively associated with ecosystems dominated by woody plants, viz. forest and scrub communities. It is true that macrofungi reach their maximum diversity and productivity in part of the forest communities, but they are important components of other communities as well. For instance, Favre (1955) recorded 202 species of macrofungi from the alpine zone (above the timber line) in the National Parc of

W. Winterhoff (ed.), Fungi in Vegetation Science, 113—149.

Switzerland only. Arnolds and De Vries (1989) calculated that at least 22% (750 species) of the macrofungi in the Netherlands (appr. 3400 species) have a preference for non-forest communities, including 365 species with an optimum in grasslands, 96 in heathlands and peat bogs, 66 in fens and mires, 50 in coastal habitats and 32 in arable fields. The total numbers of macrofungal species in grasslands and heathlands in the Netherlands surpass the corresponding numbers of phanerogams (Van der Meijden *et al.* 1983). This is often also true on the level of phytocoenoses and community-types (see § 5.4).

Mycocoenological investigations in these communities are scarce in comparison with forest communities. Studies are geographically restricted to Northwest and Central Europe. There is still a lot of basic, descriptive work to be done, before a good survey of macrofungocoenoses can be compiled and the importance of macrofungi in these ecosystems can be evaluated.

5.2. Methodological considerations

The methods of mycocoenology were discussed in a preceding chapter (2). Some aspects of the methodology deserve special attention in the communities to be considered.

1. Homogeneous phytocoenoses are in many cases too small to answer the demands of the, considerably larger, minimal area for macrofungocoenoses. In such cases it is inevitable to use fragmentary or composite plots (e.g. Winterhoff 1975, Arnolds 1981, Senn-Irlet 1987).
2. Microclimatological conditions are more variable in open communities than in forests (Barkman and Stoutjesdijk 1987). Consequently, fluctuations in species composition and sporocarp abundance are very strong as demonstrated in studies by e.g. Arnolds (1981, 1988b), Lange (1984), Brunner (1987) and Senn-Irlet (1987). See also § 5.5.2.
3. Succession of plant communities and abiotic environment is in many cases faster than in forests, especially in ephemeral communities. This may interfere with the demand for long-lasting research in stable environments (e.g. Lange 1982, Arnolds 1988b).
4. It may be difficult to distinguish between proper fungi, belonging to the investigated community, and alien fungi, associated with organisms occurring only accidentally in or near the community studied (§ 2.4). For examples, see § 5.4.1 and 5.3.5.

5.3. Ecological functions of macrofungi in plant communities outside forests

5.3.1. *Niches, substrates and exploitation of carbon sources*

The ecological niche of a macrofungal species is mainly determined by its capacity to colonize particular substrates and by its strategies to obtain carbon and nutrients. Groups of species can be established on the basis of common substrate preference or a common way of habitat exploitation. A combination of these characteristics is used in the synusial approach (Barkman 1976) and the niche-substrate groups, distinguished by Arnolds (1988a) (see § 2.4 and 5.3.5).

The most important substrates for macrofungi outside forests are (parts of) phanerogams and their remains in different stages of decay, bryophytes and their remains, amorphous organic compounds ('humus') and dung of grazing mammals. The last-mentioned coprophytic fungi occupy a special microhabitat and are discussed in chapter 6, together with macrofungi on arthropods, corpses and wood (i.c. twigs of dwarf shrubs), which are quantitatively less important.

5.3.2. *Parasitic and saprotrophic macrofungi on phanerogams and their remains*

Biotrophic parasites are frequent in many non-forest communities, but most of them belong to taxonomic groups which are not assigned to the macrofungi, e.g. Uredinales and Ustilaginales. The few exceptions are mainly inoperculate Ascomycetes of the Sclerotiniaceae and some species of *Typhula* (Basidiomycetes). Some agarics may be weak parasites, for instance *Marasmius androsaceus* (L.: Fr.) Fr. on twigs of Ericaceae (Gimingham 1972) and *M. graminum* (Lib.) Berk. on grass culms.

Saprophytic macrofungi are abundant in many of the considered communities. However, the first colonizers of leaves and stems are microfungi, yeasts and bacteria (e.g. Bell 1974). Relatively few macrofungi grow on intact standing stems and leaves of herbaceous plants, mainly species of the agaric genera *Marasmius*, *Mycena*, *Psathyrella* and *Coprinus* and small inoperculate discomycetes of the Helotiales. They are more important in swamp communities with large helophytes, such as *Typha*, *Phragmites* and large *Carices*. Many species of this category are more or less host specific.

Herbaceous litter is colonized by many species of macrofungi, for instance belonging to the agaric genera *Mycena*, *Marasmius*, *Clitocybe* and *Galerina*.

Sporocarps are under field conditions more frequent on firm components with a high lignin content (e.g. heather litter) than on soft leaves (Arnolds 1981). They are rarely host specific and often also widespread in forest communities. However, a majority of macrofungi in grasslands forms sporocarps on the soil without visible connection with a particular organic soil fraction, for instance most species of gasteromycetes, clavarioid fungi, the agaric genera *Hygrocybe*, *Camarophyllus*, *Entoloma*, *Dermoloma*, *Calocybe* and the ascomycete genus *Geoglossum*. Many of them have a distinct preference for specific plant communities outside forests. In fact their carbon source is unknown. It is usually assumed that they are involved in the decomposition of the more stable organic soil fraction, united in the term humus (Arnolds 1981). Some authors suggest a mutualistic relationship with grassland plants (e.g. Kreisel 1987, for *Hygrocybe*), although no mycorrhizal roots with basidiomycete mycelium have been described from these habitats so far. The supposition is mainly based on difficulties in isolation and subsequent culturing of these fungi in the laboratory, which they have in common with many ectomycorrhizal fungi.

In fact, surprisingly little experimental evidence exists on the role of macrofungi in the decomposition of plant remains in non-forest ecosystems. The succession of decomposers on various substrates has been studied by incubation techniques in vitro, for instance of the grass *Dactylis glomerata* by Webster (1956, 1957) and Webster and Dix (1960), litter of the herbaceous *Heracleum sphondylium* and *Urtica dioica* by Hudson (1968) and cereal straw by Magan (1988) (see survey by Bell 1974). A common characteristic of these studies is that basidiomycetes are only exceptionally isolated from these substrates, whereas they may be frequent on litter of these plants in the field.

Similar results were obtained with isolations of microorganisms from soils in grasslands (e.g. Warcup 1951a, Apinis 1958, 1970) and heathlands (e.g. Sewell 1959a,b). Both Warcup and Apinis mentioned the occurrence of many sporocarps of basidiomycetous macrofungi, but they did not succeed in isolating them with standard methods. Warcup (1951b) isolated some grassland agarics on soil plates prepared from soils collected below sporocarps, viz. of *Marasmius oreades* (Bolt.: Fr.) Fr., *Lepista nuda* (Bull.: Fr.) Cooke and *Agaricus arvensis* Schaeff., but many other fungi (e.g. *Hygrocybe* spp., *Entoloma* spp.) could not be isolated in this way. Warcup and Talbot (1962) were able to obtain sporocarps in culture of 20 species of basidiomycetes, isolated from mycelia, rhizomorphs and sclerotia in wheat fields and grasslands near Adelaide (Australia). It is remarkable that 5 species were previously undescribed and that 10 species belonged to the resupinate

Aphyllophorates and Heterobasidiomycetes, groups that are hardly reported from these habitats by taxonomists and mycocoenologists.

On the other hand, decomposition studies by pure-culture methods have demonstrated that higher basidiomycetes are in vitro effective decomposers of various herbaceous substrates, for instance various *Mycena* species on litter of e.g. *Aegopodium podagraria*, *Calamagrostis arundinacea*, *Calluna vulgaris*, *Deschampsia flexuosa* and *Filipendula ulmaria* (Hintikka 1960) and *Marasmius graminum* and *M. oreades* on litter of *Glyceria maxima* (Lindeberg 1944). It is remarkable that in these experiments, by excluding competition with other microorganisms, agarics appear to be efficient decomposers of substrates on which they are never found in the field (Arnolds 1982), for instance *Mycena sanguinolenta* (A. & S.: Fr.) Kumm. on *Aegopodium* (Hintikka 1960) and *Marasmius oreades* on *Glyceria maxima* (Lindeberg 1944).

In view of these incomplete and partially contrasting results of microbiologists, experimental mycologists and mycocoenologists, it is at present impossible to evaluate the true importance of macrofungi as decomposers in non-forest communities. Because it is generally accepted that basidiomycetes are the most important lignin decomposers (e.g. Frankland *et al.* 1982), and because lignin is a major component of many herbaceous plants (e.g. Swift *et al.* 1979), it seems necessary that these fungi (in majority macrofungi) are essential links in the decomposition process. The quantification of their role and the study of their exact substrate preference and exploitation under field conditions are practically unexplored and promising subjects of future research.

A striking feature of some saprophytic grassland fungi is their fruiting in circular patterns (fairy rings). Morphological and functional aspects of fairy rings were discussed by e.g. Kreisel and Ritter (1985), Grunda (1976) and Guminska (1976). The extension of fairy rings gives indications on the growth rate and potential age of mycelia of (part of the) saprophytic fungi. The annual extension varies from 0.1 to 0.8 metres, dependent on species and weather conditions; the radius can reach 300 metres. The maximum age of the mycelia has been calculated for some species at 160—700 years (Kreisel and Ritter 1985).

5.3.3. *Mycorrhizal macrofungi*

Mutualistic relations between green plants and fungi are frequent and important both in and outside forest communities. However, most herbs and part of the dwarf shrubs form under field conditions vesicular-arbuscular mycorrhizae with hypogeous microfungi belonging to the Endogonaceae (Phycomycetes) (e.g. Harley and Smith 1983, Sanders *et al.* 1975), which fall outside the scope of this chapter (see chapter 7). Dwarfshrubs of the Ericaceae are important components of heathland and bog communities, which are obligate mycorrhizal plants. They form various kinds of mycorrhiza (e.g. Largent *et al.* 1980), the most important one being ericoid mycorrhiza (e.g. Harley and Smith 1983). The fungal components of this type of mycorrhiza are still insufficiently known, but it was demonstrated that the ascomycete *Pezizella ericae* D. Read is able to form ericoid mycorrhiza with e.g. *Calluna* in vitro (Read 1974). Curiously enough ascocarps have never been observed in the field. Seviour *et al.* (1973) argued that Ericaceae are able to form mycorrhizae with the basidiomycetous genus *Clavaria*. Indeed sporocarps of *Clavaria argillacea* Pers.: Fr. are often found together with *Calluna* (e.g. Kramer 1969, Pirk and Tüxen 1957).

Ectomycorrhizal fungi are mainly found in association with larger woody plants, but they are also important components of macrofungocoenoses in some arctic and alpine communities with dominance of *Dryas* or dwarf species of *Betula* and *Salix* (see § 5.6.5). Some fungal species are also able to form ectomycorrhizas with the rhizomatous perennial *Polygonum viviparum* (Fontana 1977), occurring in (sub)arctic and (sub)alpine grasslands. Ectomycorrhizal symbioses are rare at low altitudes outside forest and scrub communities. An exception is formed by communities with *Salix repens*, which is associated with many species of Agaricales (Høiland and Elven 1980, Courtecuisse 1986). These communities can be in this respect equivalent to alpine dwarf willow communities, but they have a much more local distribution. In addition, some macrofungi appear to be able to live in ectomycorrhizal association with *Helianthemum* in dry grasslands (Fontana and Giovanetti 1979, Fellner and Biber 1988). This may explain earlier records of some obligate ectomycorrhizal fungi from limestone grasslands (Wilkins and Patrick 1939). Kreisel (1987) indicated species of *Hygrocybe* as possible symbiotic fungi, but there are at present hardly any arguments to support this hypothesis (see § 5.3.2).

A mutualistic relation of a different kind exists between some omphalioid Agaricales and algae forming basidiolichens (Heikkilä and Kallio 1966, 1969), which mainly grow in (sub)alpine and (sub)arctic habitats.

Besides the proper ectomycorrhizal fungi, often alien (§ 2.4) ectomycorrhizal fungi are reported from grass- and heathlands, associated with accidental woody plants inside or near the investigated plots. These data may hamper the interpretation and comparison of published data (see § 5.4.1).

5.3.4. *Bryophytic macrofungi*

Many species of Agaricales (e.g. in the genera *Galerina, Omphalina, Rickenella, Psilocybe*), clavarioid Aphyllophorales (e.g. *Clavaria, Clavulinopsis*), Pezizales (e.g. *Octospora, Lamprospora*) and Helotiales (e.g. *Geoglossum*) are predominantly or exclusively found together with bryophytes (see also paragraph 6.6). They were indicated as bryophytic by Arnolds (1982). The relationships between macrofungi and bryophytes are still insufficiently known; they may be direct as parasites, saprophytes or symbionts, or indirect due to a favourable microclimate in moss carpets for sporocarp formation. Benkert (1976) enumerated the associations between Pezizales and bryophytes, which appear to be often host-specific. Döbbeler (1979) demonstrated, that many Ascomycetes are connected with moss plants by characteristic appressoria and that they are able to penetrate living cells with haustoria, hence that they are parasites. A parasitic relation between *Sphagnum squarrosum* and the inoperculate *Discinella schimperi* (Nawaschin) was observed by Redhead and Spicer (1981). Redhead (1981) established in laboratory experiments that the agaric *Tephrocybe palustris* (Peck) Donk is able to penetrate and kill healthy *Sphagnum* plants. This is confirmed by observations in the field (e.g. Watling 1978). The mycelium of *Galerina paludosa* (Fr.) Kühner forms in vitro haustoria in living cells of *Sphagnum* rhizoids, without visual damage to the plant. Similar structures were observed in the field in *Pleurozium schreberi* (Brid.) Mitt., associated with *Rickenella fibula* (Bull.: Fr.) Raith. (Redhead 1981, Kost 1984). Redhead considered these associations as 'balanced parasitism', but a kind of mutualistic relationship cannot be excluded on forehand (Senn-Irlet 1987).

It has been demonstrated that several agarics are effective decomposers of bryophyte litter in vitro, for instance *Collybia peronata* (Bolt.: Fr.) Sing., *Mycena clavicularis* (Fr.) Gill. (Mikola 1956), *M. galopus* (Pers.: Fr.) Kumm., *M. sanguinolenta* (A. & S.: Fr.) Kumm., *M. vitilis* (Fr.) Quél. (under field conditions never on mosses, but usually on twigs!) and *M. filopes* (Bull.: Fr.) Kumm. (Hintikka 1960). The decomposition rate of pleurocarpous mosses is comparable with that of deciduous leaves, but much higher than that of coniferous litter and *Sphagnum*.

5.3.5. *Niche-substrate groups in different plant communities*

Niche-substrate groups of macrofungi are based on common properties concerning substrate preference and function (way of habitat exploitation) (Arnolds 1988a; § 2.4). These groups can be distinguished at several levels of integration. An example of a detailed approach is given in Table I. It elucidates the important difference between proper fungi, growing on substrates produced in and belonging to the community investigated, and alien fungi, growing on substrates produced outside the community or only accidentally found within the community. The most important group of alien fungi consist of ectomycorrhizal species, associated with woody plants which grow accidentally in or near the community concerned, for instance young trees in heathland (see also examples in § 5.4.1). The distinction between proper and alien fungi is sometimes complicated: Among the coprophytic fungi in grass heath the species on rabbit pellets are considered as proper, species on horse dung as alien. On the other hand, sporocarps of *Marasmius androsaceus* have to be regarded as proper to the community when growing on heather and grass litter, but as alien on blow-in oak leaves. The position of the fungi on arthropods and fox dung is doubtful in this respect.

The distribution of species among the niche-substrate groups provides information on the ecological conditions for macrofungi in different plant communities. Spectra of main groups and sub-groups of terrestrial fungi in grass- and heathlands in the Netherlands were published by Arnolds (1981). Examples are presented in Table II, together with spectra of some alpine communities, after Senn-Irlet (1987). It is evident that (proper) ectomycorrhizal fungi are only important in alpine communities with ectomycorrhizal dwarf shrubs. Litter saprophytes are dominant in heathland communities (Ericion) with slowly decaying plant remains, whereas humus saprophytes are more important in the presented grassland types. Bryophytic fungi are important in dry, poor grasslands of the Thero-Airion, and their share decreases strongly on richer soils, together with the abundance of bryophytes, being absent from the strongly fertilized Poo-Lolietum. The different proportions of parasitic and saprophytic fungi on bryophytes in alpine and lowland communities are mainly due to a different estimation of the functional relationships by Senn-Irlet (1987) and Arnolds (1981), respectively. Subcoprophytic fungi on enriched spots are frequent in the Poo-Lolietum. The slightly fertilized, wet hayfields of the Calthion take an intermediate position.

5.4. Species diversity, abundance and productivity of sporocarps

5.4.1. *Species diversity of macrofungi in various plant communities*

A selection of data on the number of species, abundance of sporocarps and sporocarp productivity of macrofungi in the considered plant communities is summarized in Tables III—IV. In fact only a limited number of accurate data

Table I. Niche-substrate groups of macrofungi in dry grass heaths in the Netherlands (after Arnolds 1988)

I Proper niche-substrate groups

1. On litter of *Molinia caerulea* (s): *Mycena adonis* group. Ch.: *Mycena adonis*. Others: *M. epipterygia* and some other species of group 2.
2. On litter of *Festuca, Nardus* and other herbs and grasses (s): *Mycena pelliculosa* group. Ch.: *M. pelliculosa, Galerina allospora*. Others: *Mycena sanguinolenta, Marasmius androsaceus, Clitocybe vibecina* etc.
3. On twigs of *Calluna* and *Erica* (s or p?): *Marasmius androsaceus* group. Ch.: *Athelia epiphylla*. Others: *M. androsaceus* (p?).
4. On litter of *Calluna* and *Erica* (s): *Mycena galopus*-group. Ch.: *Mycena galopus, M. cinerella, Collybia dryophila*. Others: *Marasmius androsaceus, Mycena sanguinolenta, Clitocybe vibecina*.
5. On acrocarp mosses (s): *Rickenella fibula* group. Ch.: *Rickenella fibula*. Others: *Galerina hypnorum, G. vittaeformis* ssp. *atkinsoniana*.
6. On pleurocarp mosses (s): *Galerina pumila* group. Ch.: *G. pumila, G. hypnorum, G. vittaeformis* ssp. *atkinsoniana*.
7. On humus in soil (s): *Mycena leptocephala, M. filopes, M. sepia, Entoloma vinaceum, E. conferendum* etc.
8. On moss, grass or humus near excrements or urine-saturated places (s): *Tephrocybe tesquorum* group. Ch.: *Tephrocybe tesquorum, Panaeolus fimicola, Agrocybe semiorbicularis* s.l., *Clitopilus hobsonii*. Others: *Sphaerobolus stellatus*.
9. On rabbit pellets (s): *Psilocybe coprophila* group. Ch.: *Psilocybe coprophila, Lasiobolus ciliatus, Coprinus miser*. Others: *Sphaerobolus stellatus*.
10. On dead carpophores of agarics (s): *Collybia cirrhata* group. Ch.: *Collybia cirrhata*.
11. On caterpillars and pupae of Lepidoptera (p): *Paecilomyces farinosus*-group. Ch.: *Paecilomyces farinosus*.

II. Alien niche-substrate groups

12. On blown-in oak leaves (s): *Mycena polyadelpha* group. Ch.: *Mycena polyadelpha*. Others: *Marasmius androsaceus*.
13. Associated with *Quercus robur* (m): *Cortinarius anomalus*-group. Ch.: *Cortinarius anomalus*. Others: *Laccaria proxima, Thelephora terrestris* (or s?, then group 2).
14. Associated with *Betula* spec. (m): *Paxillus involutus*-group. Ch.: *Paxillus involutus*. Others: *Laccaria proxima*.
15. On excrements of horse (s): *Coprinus curtus*-group. Ch.: *Coprinus curtus*.
16. On excrements of fox (s): *Cheilymenia pulcherrima*-group. Ch.: *Cheilymenia pulcherrima*.

The data are based on the study of six plots of the sociation of *Galium saxatile* and *Ptilidium ciliare* by Arnolds (1981).
Functional groups: s = saprotrophic; m = mycorrhizal; p = parasitic. Ch = characteristic species, i.e. a species that is much more frequent in one group than in all other groups within the investigated community type. Nomenclature of fungi after Arnolds (1981).

are available. Besides, the figures are not directly comparable, due to differences in methodology (for instance size and number of plots, duration of research, frequency of visits (see § 2.5)), completeness of species lists (small sporocarps or difficult groups may have been omitted) and applied species

Table II. Spectra of niche-substrate groups of proper terrestrial macrofungi in various grass- and heathland communities in the Netherlands (after Arnolds 1981) and alpine communities in Switzerland (after Senn-Irlet 1987), based on the proportion of the average species number per plot (n = number of plots)

		% Niche-substrate group								
Plant community	n	hp	hm	hs	bp	bs	li	hu	sc	ls
alpine communities: (after Senn-Irlet 1987)										
Epilobietum fleischeri	2	3	43	1	8	12	4	31	—	—
Salicetum herbaceae	3	—	50	—	14	7	5	20	4	—
Seslerio-Caricetum sempervirentis	4	—	30	—	—	—	6	54	—	11
Lowland communities: (after Arnolds 1981)										
Ericion tetralicis	7	—	—	6	—	15	45	24	9	—
Thero-Airion	12	1	—	7	—	33	12	41	7	—
Calthion palustris	5	—	—	15	—	15	3	62	6	—
Poo-Lolietum	7	2	—	7	—	—	3	54	33	—

Niche-substrate groups (after Arnolds 1981) hp: parasitic fungi on herbs, hm = mycorrhizal fungi with dwarf shrubs, hs = saprotrophic fungi on stems and leaves, bp = parasitic fungi on bryophytes, bs = saprotrophic fungi on bryophytes, li = saprotrophic fungi on litter, hu = saprotrophic fungi on humus, sc = subcoprophytic fungi, ls = lignicolous fungi.

concepts. They give only an order of magnitude, in fact minimum values since all mycocoenological investigations are carried out with considerable intervals between visits, during which sporocarps may have remained unnoticed. Many data are obviously incomplete because they are based on only one or few visits per plot (e.g. Heinemann 1956, Krisai 1987) or on the use of very wide species concepts (e.g. Wilkins *et al.* 1939). These data sets are indicated in the Tables with an asterisk *.

The interpretation of published data is hampered by the incorporation of various proportions of alien species (§ 2.4) in the species lists, mainly of ectomycorrhizal fungi associated with (accidental) woody plants inside or near the plots. Some examples may illustrate this problem. Pirk and Tüxen (1957) observed 78 species of macrofungi in 22 plots of dry heathlands (Calluno-Genistetum), including 48 obligate or facultative ectomycorrhizal species. Among the proposed 11 character species of the Calluno-Genistetum are no less than 8 ectomycorrhizal species. Rudnicka-Jezierska (1969) listed 59 species from five plots of dune heaths, dominated by *Calluna vulgaris.* According to her vegetation relevés, only some scattered trees of *Pinus sylvestris, Betula verrucosa* and *Quercus robur* are found within the plots. Yet 39 species (66%) are presumably ectomycorrhizal and 8 species

Table III. Number of species and number and productivity of sporocarps of macrofungi in various grassland and weed communities in Western and Central Europe

Plant community (with reference)	n	su	yr	NS	amDC	aPCy	Gm	Hy	En	My	Ga	Co	Pa
A. DRY GRASSLANDS ON CALCAREOUS SOILS													
Festucetum pallentis (1)*	1	?	1	10	?	?	6	2	—	—	—	—	—
Corynephorus-Koeleria glauca comm. (2)	4	200	5	4.2	250	?	4	—	—	—	—	—	—
Koeleria glauca comm. (3)	14	150—2100	3	12	?	?	7	—	2	2	1	1	—
Koelerio-Poetum xerophilae (4)	1	1000	11	43	?	?	9	5	7	—	—	1	—
Allio-Stipetum capillatae (8*)	3	?	2	19	?	?	5	—	4	3	1	1	1
Allio-Stipetum capillatae (19)*	14	?	2—4	24	?	?	14	3	8	6	1	4	—
Festuca lemani comm. typ. (3)	13	100—300	3	28	?	?	13	—	7	8	1	6	2
Poa angustifolia comm. (17)*	1	?	2	45	?	?	5	—	2	5	2	2	2
Xerobrometum (5)	1	7500	3	76	?	?	6	6	12	3	2	1	5
Xerobromion/Mesobromion (6)*	1	25,000	4	135	?	?	9	8	28	3	4	5	4
Adonido-Brachypodietum (18)*	3	?	2	28	?	?	6	1	10	5	3	1	1
Genistello-Phleetum (19)*	14	?	2—4	14	?	?	13	3	6	6	3	2	1
Seselio-Mesobrometum (7)	2	800—900	3	37	?	?	—	5	1	10	2	5	—
Dauco-Mesobrometum (7)	1	1000	3	39	?	?	—	2	2	2	2	5	—
Medicagini-Mesobrometum (7)	1	1000	3	39	?	?	1	4	1	3	1	3	2
Origano-Brachypodietum (1)*	1	?	1	23	?	?	8	4	—	—	—	—	—
Mesobromion (8)	7	?	2	53	?	?	5	13	5	5	1	1	3
Anthyllido-Trifolietum (9)	3	600	2	26	1300	?	2	14	5	7	1	1	1
Koelerio-Gentianetum (10)*	1	9000	3	94	?	?	7	22	8	7	1	6	2
'Trockenrasen' (11)*	1	> 10,000	2	91	?	?	8	18	4	2	1	3	4
Lawn on chalk (12)	1	380	7	63	9500	?	4	5	7	6	1	2	2
B. UNFERTILIZED GRASSLANDS ON NOT CALCAREOUS SOILS													
Spergulo-Corynephoretum (13)	5	100—300	3	8	500	0.1	1	—	2	3	2	—	—
Thero-Airion (13)	12	100—400	3	35	5000	2.2	9	9	26	16	9	5	3
Grasslands on sand (8)	7	?	2	50	?	?	4	13	5	5	1	1	3
Potentillo-Festucetum (4)	1	1000	11	61	?	?	1	10	16	1	1	2	1
Nardetum strictae (9)	3	600	2	25	3450	?	3	10	5	6	1	1	1
Cynosuro-Lolietum luzuletosum (14)	2	2500	3	75	?	?	4	11	10	5	2	5	6

Table III (continued)

Plant community (with reference)	n	su	yr	NS	amDC	aPCy	Gm	Hy	En	My	Ga	Co	Pa
Junco-Molinion (13)	5	150—600	3	18	450	0.1	—	3	10	7	8	1	3
Calthion palustris (13)	5	300—500	3	17	550	0.1	—	1	11	6	8	1	3
Stellario-Deschampsietum (15)*	1	75	1	22	3300	0.3	—	1	6	2	5	—	—
C. FERTILIZED GRASSLANDS													
Arrhenatheretum (16)*	1	50	1	16	6100	5.0	1	1	—	4	1	1	1
Arrhenatheretum (1)*	1	?	1	21	?	?	1	1	—	—	1	3	2
Lolio-Cynosuretum (1)*	1	?	1	25	?	?	6	3	1	—	3	3	1
Lolio-Cynosuretum (13)	9	400—1000	3	21	650	1.0	4	2	9	10	7	6	5
Grasslands on clay (8)	6	?	2	29	?	?	1	7	2	2	1	1	2
Poo-Lolietum (13)	7	500	3	19	1000	1.3	3	—	2	5	1	8	4
D. WEED COMMUNITIES													
Vicietum tetraspermae (1)*	1	?	1	7	?	?	1	—	—	—	—	—	1
Lamio-Veronicetum politae (1)*	1	?	1	6	?	?	1	—	—	—	—	—	1
Euphorbio-Melandrietum, intensive (16)	22	?	3	4.7	?	?	1	—	—	—	—	11	2
Euphorbio-Melandrietum, extensive (16)	4	?	3	21	?	?	1	—	—	—	—	11	2
Medicago sativa fields (16)	4	?	3	9.2	?	?	1	—	—	—	—	3	1
Bromus-Eragrostis comm. (17)*	1	?	2	2	?	?	2	—	—	—	—	—	—

n = number of studied plots. — su: surface (m^2). — yr = number of years of observation. — NS = average number of proper species of macrofungi per plot. — amDC = average maximum density of sporocarps per year. — aPCy = average prodcutivity of sporocarps per year (in kg dr.wt/ha). — Gm = Number of species of Gasteromycetes in all studied plots. — Hy = Idem for the genus *Hygrocybe* (incl. *Camarophyllus*). — En = Idem for *Entoloma*. — My = Idem for *Mycena*. — Ga = Idem for *Galerina*. — Co = Idem for *Conocybe*. — Pa = Idem for *Panaeolus*.

References (for complete quotations, see References at end of paper; data marked with * are incomplete or based on deviating methods): 1: Poland, Ojcow National Park; Wojewoda 1975. 2: Poland, Kampinos National Park; Rudnicka-Jezierska 1969. 3: GFR, Baden-Württenberg, Sandhausen; Winterhoff 1975. 4: Switzerland, Unterengadin (1100—1200 m); Horak 1985. 5: GFR, Bayern, Augsburg; Stangl 1970. 6: GFR, Bayern, Garchinger Heide; Einhellinger 1969. 7: Switzerland, Schaffhauser Jura (1200 m); Brunner 1987. 8: Great-Britain, Oxford, Leicester, Haslemere; Wilkins and Patrick 1939. 9: Poland, Pieniny National Park (950 m); Guminska 1976. 10: Netherlands, Limburg, Bemelerberg (200 m); Kuyper and Schreurs 1984. 11: GFR, Schleswig-Holstein, Lübeck; Glowinski 1984. 12: Great-Britain, Bristol; Bond 1981. 13: Netherlands, Drenthe (20m); Arnolds 1981. 14: GFR, Hamburg (10 m); Neuhoff 1949, 1950. 15: Poland, Kampinos National Park; Sadowska 1974. 16: GDR, Halle; Hille 1983. 17: GFR, Baden-Württenberg, Schwetzingen; Winterhoff 1983. 18: GFR, Bayern, gypsum-hills; Winterhoff 1986. 19: GFR, Baden-Württenberg, Hessen, Oberrheinebene; Winterhoff 1978.

Table IV. Number of species and number and productivity of sporocarps of macrofungi in various alpine, heathland and bog communities in Western and Central Europe

Plant community (with reference)	n	Su	yr	NS	amDC	aPCy	Gm	Hy	En	My	Ga	Ct	In	%em
A. ALPINE AND ARCTIC COMMUNITIES														
Salicetum herbaceae (1)	4	30—60	3	8	3450	0.9	—	—	5	—	2	4	4	52
Salix herbacea comm. (2)*	10	1	1	3.5	?	?	—	—	1	3	3	3	1	61
Salix herbacea comm. (3)	2	100	7	?	?	0.4	—	1	4	1	1	4	1	45
Salicetum retuso-reticulatae (1)	3	30—100	3	29	1100	1.7	1	1	9	4	6	10	12	42
Betula nana heath (2)*	3	25	1	7.7	?	?	1	—	1	—	1	3	4	62
Betula nana health (3)	4	100	7	?	?	6.5	—	—	2	3	1	11	5	57
Betula glandulosa heath (2)*	10	1	1	3.9	?	?	—	1	—	2	2	2	1	47
Seslerio-Caricetum with Dryas (1)	4	30—120	3	9	400	0.6	3	3	8	1	—	2	6	33
Saxifrago-Caricetum frigidae & al. (1)	5	70—100	3	23	8000	1.1	—	1	6	3	6	12	10	46
Epilobietum fleischeri (1)	2	50—60	3	36	28,400	16.0	1	3	4	1	3	4	10	39
Caricetum curvulae (1)	1	100	3	3	150	0.1	—	—	—	—	—	—	—	—
B. HEATHLANDS AND GRASS HEATHS														
Carex ericetorum-Calluna comm. (4)	4	200	5	4.8	300	?	4	—	1	—	—	—	—	—
Calluno-Genistion (5, 9)	29	100	2	15	?	?	—	—	7	8	9	—	—	—
Calluno-Genistion (6)*	29	20—1000	1	3.1	2	?	—	1	3	5	2	—	—	—
Calluno-Genistetum typicum (7)	1	100	2	14	8700	?	—	—	2	4	3	—	—	—
Calluno-Genistetum typicum (8)*	25	100	1	3.4	?	?	2	1	2	1	2	—	—	—
Violion caninae (9)	7	80—400	3	21	2500	0.6	2	2	11	10	9	—	—	—
Ericetum tetralicis (9)	7	400—600	3	18	1400	0.4	2	2	9	9	8	—	—	—
Nardo-Juncion squarrosae (9)	5	150—800	3	28	1900	0.3	1	3	11	13	11	—	—	—
Molinia sociation (10)*	10	1	3	8.3	?	?	—	—	4	6	3	—	—	—

Table IV (continued)

Plant community (with reference)	n	Su	yr	NS	amDC	aPCy	Gm	Hy	En	My	Ga	Ct	In	%em
C. BOGS AND MARSHES														
Sphagnum ass. in lagg zone (10)*	15	1	3	4.1	?	?	—	—	—	3	4	—	—	—
Sphagnetum magellanici (11)*	16	?	1—2	3.4	?	?	—	—	—	—	5	2	—	14
Sphagnum-Eriophorum ass. (10)*	10	1	3	7.0	?	?	—	—	1	4	4	—	—	—
Eriophorum vaginatum ass. (12)	3	5—14	1	3.3	5000	?	—	—	—	2	1	—	—	—
Eriophoro-Trichophoretum cespitosae (11)	12	?	1—2	2.0	?	?	—	2	—	—	2	—	—	—
Eriophorum-Empetrum ass. (10)*	10	1	3	6.8	?	?	—	—	1	7	5	—	—	—
Vaccinium-Empetrum ass. (10)*	10	1	3	8.8	?	?	—	—	1	7	3	—	—	—
Caricetum limosae (11)*	4	?	1—2	3.7	?	?	—	—	—	—	2	—	—	—
Caricetum lasiocarpae (11)*	3	?	1	1.7	?	?	—	—	—	—	1	—	—	—
Caricetum rostratae (11)*	7	?	1—2	1.4	?	?	—	2	—	—	1	—	—	—
Caricetum davallianae (11)*	1	?	1	2	?	?	—	1	—	—	—	—	—	—
Caricetum nigrae (11)*	1	?	1	1	?	?	—	—	—	1	—	—	—	—
Caricetum elatae (13)	1	25	1	4	160	<0.1	—	—	—	—	—	—	—	—

— n = number of studied plots. — su = surface (m^2). — yr = number of years of observation. — NS = average number of proper species of macrofungi per plot. — amDC = average maximum density of sporocarps per year. — aPCy = average productivity of sporocarps per year (in kg dr.wt/ha). — Gm = Number of species of Gasteromycetes in all studied plots. — Hy = Idem for the genus *Hygrocybe* (incl. *Camarophyllus*). — En = Idem for *Entoloma*. — My = Idem for *Mycena*. — Ga = Idem for *Galerina*. — Ct = Idem for *Cortinarius* (incl. *Dermocybe*). — In = Idem for *Inocybe*. — %em = % proportion of total number of species, forming ectomycorrhiza.

References (for complete quotations, see References at end of paper; data marked with * are incomplete or based on deviating methods): 1: Switzerland, Central Alps (1900—2500 m); Senn-Irlet 1987. 2: Greenland; Lange 1957. 3: Finland, northwest Lapland, Kilpisjärvi (630 m); Metsänheimo 1987. 4: Poland, Kampinos National Park; Rudnicka-Jezierska 1969. 5: Netherlands, Drenthe (20 m); Kramer 1969. 6: Belgium, Brussels (100 m); Heinemann 1956. 7: GFR, Nordrhein-Westfalen, Münster; Runge 1960. 8: GFR, Nordrhein-Westfalen; Pirk and Tüxen 1957. 9: Netherlands, Drenthe (20 m); Arnolds 1981. 10: Denmark, Maglemoose; Lange 1948. 11: Austria, Oberen Murtal (1200—1700 m); Krisai 1987. 12: GFR, Nordrhein-Westfalen, Emsdetter Venn; Augustin and Runge 1968. 13: Poland, Kampinos National Park; Sadowska 1974.

(14%) wood-inhabiting, leaving only 12 proper saprophytic fungi (20%). Sadowska (1974) analyzed the productivity of fungi in a moist hayfield near a forest edge. Ectomycorrhizal fungi contributed 22% to the number of species (36), 38% to the number of sporocarps (407) and 71% to the dry weight (22 g/100 m^2). Alien fungi have been, as far as possible, omitted from the data calculated in the Tables III—IV.

A comparison of representative mycocoenological plots (plots ± 100—1000 m^2, studied during at least 2 years) shows that species diversity varies enormously between less than 1 to approximately 50 per plot. Mycocoenoses rich in species (> 30) are mainly found in poor, not or weakly fertilized grasslands on mesic to dry, weakly acid to basic, undisturbed soil and also in wet, alpine bank communities (Epilobietum fleischeri, Senn-Irlet 1987) deviating from the mentioned vegetation types in the presence of ectomycorrhizal fungi. Moderately rich in macrofungi (15—30 species) are strongly fertilized grasslands, unfertilized grasslands on wet soils, most heathlands and snowbed communities on calcareous soils. Poor in species (< 15 species) are dry alpine meadows, snowbed communities on acid soils, many bog and marsh communities, pioneer communities on dry sand dunes and ephemeral weed communities. The limiting conditions for macrofungi are different in each of these communities.

Arnolds (1981) investigated the correlation between the number of species of macrofungi and phanerogams in 64 plots in grass- and heathlands in the Netherlands (Figure 1). Senn-Irlet (1987) carried out a similar study in 21 plots in alpine communities (Figure 2). Both authors found only a weak, positive correlation and large differences between various plant communities. Some are poor in both phanerogams and macrofungi (Spergulo-Corynephorion, Salicetum herbaceae), others are rich in phanerogams, but poor in macrofungi (e.g. wet grasslands of Calthion and Junco-Molinion, dry alpine grasslands of Seslerio-Caricetum sempervirentis), or rich in both groups of organisms (dry grasslands of Thero-Airion, alpine seapage communities of Saxifrago-Caricetum frigidae and related associations). Apparently the diversity of phanerogams, and consequently of potential host plants and litter types, is not the most important factor for the diversity of macrofungocoenoses. Edaphic and climatological conditions appear to be more important (§ 5.5).

5.4.2. *Abundance and productivity of sporocarps*

Data on abundance, and especially on productivity of sporocarps in the plant communities concerned are scarce and even more influenced by methodo-

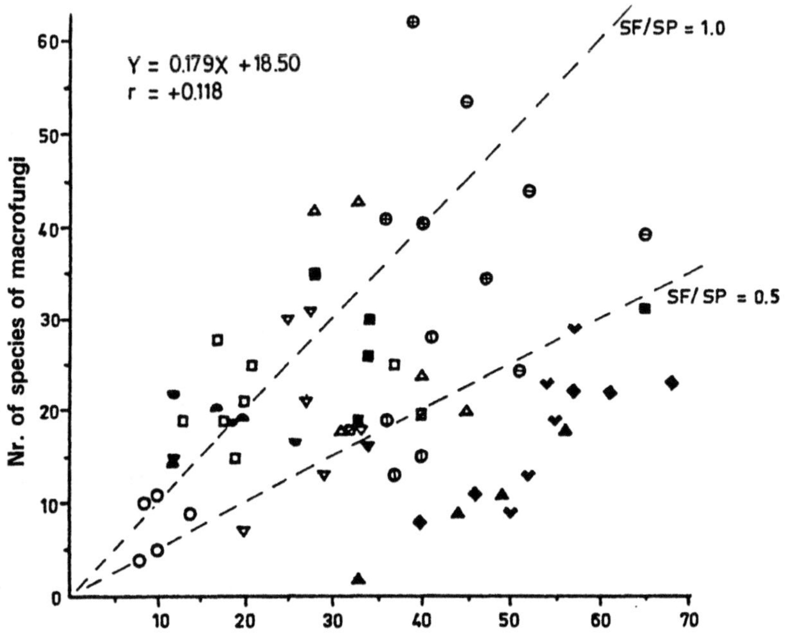

Fig. 1. The relation between the numbers of species of proper terrestrial macrofungi and phanerogamic plants in 64 plots in grass- and heathlands in the Netherlands. The symbols indicate different plant communities; for an explanation see Figure 3 (after Arnolds 1981).

logical factors than data on species diversity. This may explain part of the large variation in figures from related plant communities (Table III—IV). The maximum annual sporocarp density varies from less than 100 to 9,000 (−28,000) per 1000 m^2. A significant, positive correlation exists between the species diversity and sporocarp abundance in both lowland grasslands (Arnolds 1981: 147) and alpine communities (Senn-Irlet 1987: 82), but some community types are deviating in this respect. For instance, moderately fertilized meadows of the Lolio-Cynosuretum are relatively rich in species, but poor in sporocarps and dry grass heaths of the Violion caninae show an opposite relation (Arnolds 1981). Alpine snowbed communities on calcareous soil (Salicetum retuso-reticulatae) are very rich in species, but remarkably poor in sporocarps (Senn-Irlet 1987).

The average sporocarp productivity varies in most grass- and heathland communities between 0.1 and 2 kg dr wt/ha/yr, with some extreme values up to 16 kg (Tables III—IV). The produced fresh weight is approximately a factor 10 higher. Arnolds (1981: 146) demonstrated a positive, but rather weak correlation between sporocarp density and productivity in grass- and

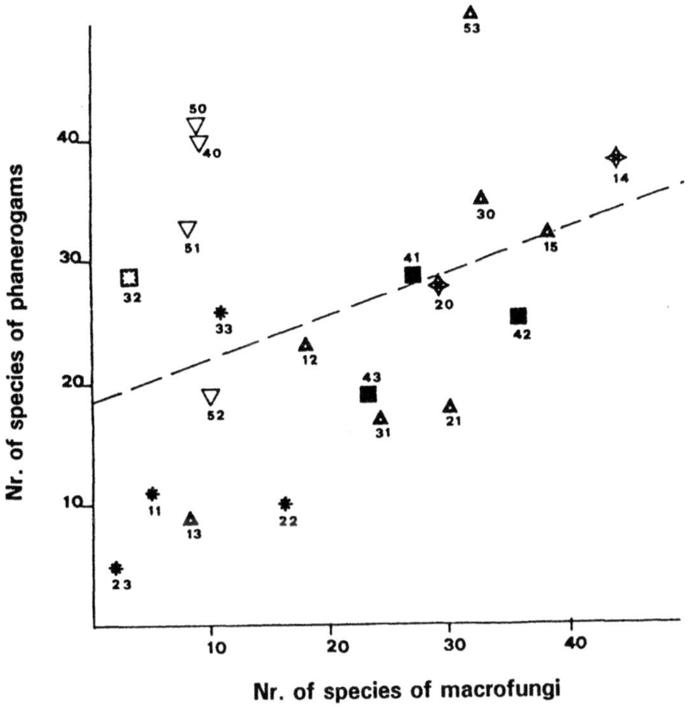

Fig. 2. The relation between the numbers of species of phanerogamic plants and macrofungi in 21 plots in alpine communities in Switzerland. The symbols indicate different plant communities; for an explanation see Figure 4 (after Senn-Irlet 1987).

heathlands in the Netherlands. The weak character of this correlation is caused by great differences in specific weight between species. For instance, dry basidiocarps of *Mycena pudica* Hora weight less than 0.1 mg, of *Macrolepiota procera* (Scop.: Fr.) Sing. 13 g (Arnolds 1981: 270). Consequently, communities with a high proportion of heavy sporocarps have a relatively high productivity (e.g. moderately fertilized meadows of the Lolio-Cynosuretum with *Marasmius oreades* and *Clitocybe* spp. as dominant fungi) and communities with many tiny sporocarps have a relatively low productivity (e.g. dry grass heaths, dominated by *Mycena* and *Galerina* spp.; Arnolds 1981).

5.5. Relations between macrofungocoenoses and environmental conditions

5.5.1. *Soil conditions*

Correlations between characteristics of macrofungocoenoses and soil properties have been studied by a number of mycocoenologists. Relatively extensive information was published by Senn-Irlet (1987) for macrofungocoenoses in alpine communities and by Arnolds (1981, 1982) and Brunner (1987) for grass- and heathlands. Rough conclusions can be drawn from the data presented in Tables III—IV.

Hydrological conditions are very important for macrofungocoenoses. Richest in both species and sporocarps are mesic to dry habitats. Both extremely dry environments and permanently water-logged substrates are unfavourable, although they accommodate a number of specialized fungi. The poor macrofungal flora on most very wet soils is apparently caused by shortage of oxygen, because some plant communities under the influence of streaming water appear to be rich in macrofungi (Senn-Irlet 1987). Modess (1941) showed in experiments that insufficient aeration inhibits the growth of basidiomycetes and ligninolysis is an oxygen-dependent process.

An at first sight remarkable relation exists between organic matter content of the soil and sporocarp productivity. The highest productivity was measured by Arnolds (1981) in grasslands on soils with a (rather) low organic matter content between 2 and 10%. The productivity in grasslands on soils with a higher organic matter is considerably lower (Figure 3). Senn-Irlet (1987) found a similar correlation in alpine communities (Figure 4). These results suggest that a low activity of macrofungi may cause the accumulation of a stable organic soil fraction, either stable humus or peat, in particular in wet environments where anaerobic conditions are prevailing.

No general correlation exists between characteristics of macrofungocoenoses and the C/N ratio of the organic matter. Arnolds (1981) found grassland plots with high and low sporocarp productivities in C/N range of 14 to 25; Senn-Irlet (1987) in alpine communities at C/N ratios between 8 and 25. The productivity was constantly low in grass heaths and heathlands with C/N ratios between 25 and 35 (Arnolds 1981). The low availability of nitrogen may influence litter quality and lead in particular to a higher ratio between lignine and cellulose, so that the decomposition process is retarded (Frankland *et al.* 1982). Arnolds (1981) found a significant positive correlation between the proportion of litter-inhabiting fungi and the C/N ratio in temperate grass- and heathlands.

The acidity of the soil is within a broad range not a prevailing factor for the development of macrofungocoenoses. Examples rich and poor in both

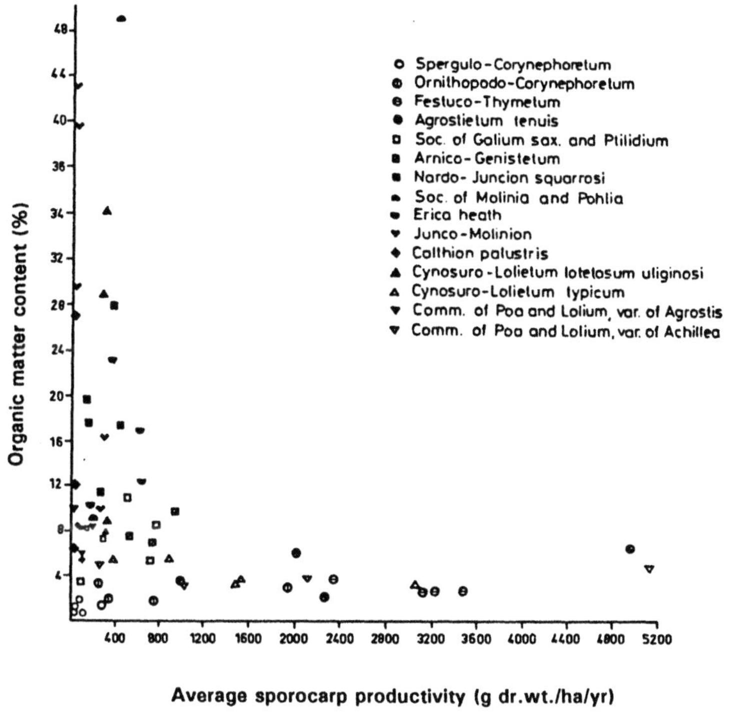

Fig. 3. The relation between the average annual sporocarp productivity of proper terrestrial fungi and the organic matter content of the top soil (0—5 cm) in 64 plots in grass- and heathlands in the Netherlands. The symbols indicate different plant communities (after Arnolds 1981).

species and sporocarps were reported from acid (pH 4.5) to basic (pH 7.5) substrates (Tables III—IV; Arnolds 1981, Senn-Irlet 1987, Brunner 1987). The productivity was comparatively low in grass heaths and grasslands with a pH between 3.5 and 4.5, which were at the same time characterized by a high C/N ratio (see above). Some species are almost indifferent to soil pH, but others are confined to a certain pH range (Bohus 1984, Arnolds 1982; Figures 5—6).

The texture of the soil seems to be important to some extent. Wilkins and Patrick (1939) observed in grasslands considerably higher numbers of sporocarps on sand than on limestone, and these were higher than on clay. Senn-Irlet (1987) found in alpine communities a positive correlation between sand content of the top soil and species diversity of macrofungi, which was ascribed to the better aeration in sandy soils. However, at the same time the

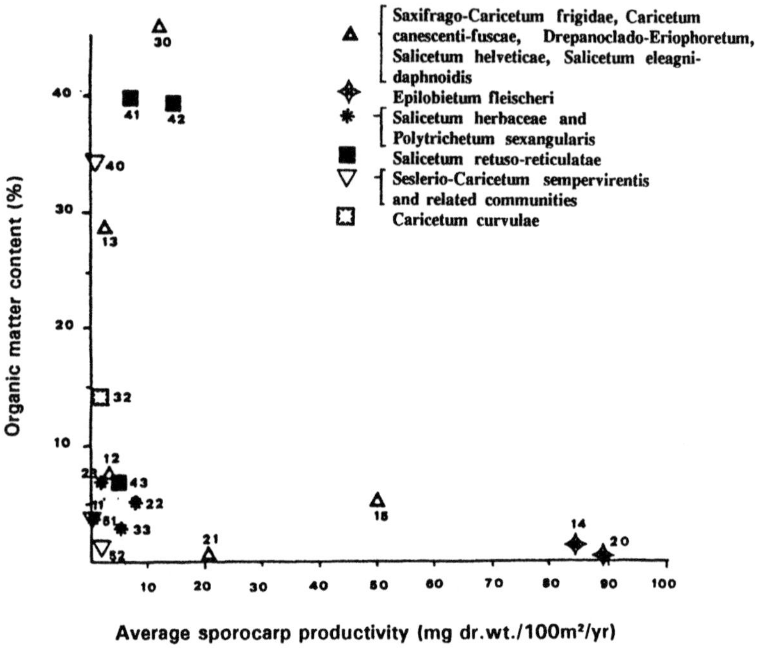

Fig. 4. The relation between the average annual sporocarp productivity of macrofungi and the organic matter content of the top soil in 21 plots in alpine communities in Switzerland. The symbols indicate different plant communities (after Senn-Irlet 1981).

availability of nitrogen and other minerals is in general lower on sand than on clay, which offers an alternative explanation (see below).

Little exact information is available on the influence of mineral content of the soil on macrofungocoenoses. It is striking that halophyte communities along the coast are very poor in macrofungi (Courtecuisse 1986), but it is less obvious whether this has to be ascribed to the chloride concentration or periodical inundations with seawater. The lime content of the soil is not correlated with species diversity or sporocarp productivity, but strongly influences the composition of fungal communities, probably by means of the pH (Tables III, IV; Senn-Irlet 1987). The influences of phosphorus and potassium are unclear. Arnolds (1981, 1982) found correlations between the abundance of various grassland species and the phosphate content of the soil, but in this case high phosphate levels were correlated with high nitrogen levels. Lange (1984) compared plots in lawns with different levels of phosphorus fertilization and found no relation with macrofungi. However, soil analyses pointed out that soils in all plots had a high phosphorus content, irrespective of the level of fertilization. The influence of nitrogen is discussed below in the context of fertilization experiments (§ 5.5.3).

The autecological relations between grassland macrofungi and soil conditions were described by e.g. Andersson (1950), Lange (1982), Arnolds (1982) and Brunner and Horak (1988). Examples of ecological spectra of two *Mycena* species in grass- and heathlands in the Netherlands are given in Figures 5 and 6.

5.5.2. *Climatological conditions*

Well-developed macrofungocoenoses outside forests are found in all climatological regions of Europe, from arctic and alpine areas to temperate and mediterranean areas. Climatological regions have their own mycoflora. For instance, arcto-alpine areas are characterized by numerous characteristic species (Favre 1955, Gulden *et al.* 1985). On the other hand, a considerable number of macrofungi occurs from alpine meadows to coastal grasslands and mediterranean dwarf scrub communities, for instance many species of *Hygrocybe*. Such an amplitude is exceptional among phanerogams.

The potential fruiting period of macrofungi is mainly determined by temperature. Mycelial growth in vitro and fruiting in most species are inhibited by temperatures below 5 °C (Fries 1949, Bohus 1957) and most sporocarps are killed at temperatures below −5 °C (Arnolds 1981: 183). Such temperatures are more easily reached in open communities than in forests. On the other hand, a few species need a stimulus by frost for sporocarp formation (Brunner 1987: 222). In view of these properties the main fruiting period is in alpine and arctic areas from July until middle September (Senn-Irlet 1987), in the lowland of Denmark from early September until late October (Lange 1948, 1984) and in the Netherlands from late September until middle November (Arnolds 1981, 1988b).

Within uniform climatological areas fruiting is influenced by microclimatological conditions. For instance, in the Netherlands a peak in fruiting is observed in November and early December in the exposed, dry sand dune community of the Spergulo-Corynephorion, but in October until early November in mesic grassland communities (Arnolds 1981). Winterhoff (1975) observed in the *Festuca lemanii* community on dry sand dunes in South Germany a higher species diversity in stands along forest margins with a more humid microclimate than in exposed stands.

Varying weather conditions are very important for the explanation of periodicity and fluctuations in fruiting. Several studies pay special attention to the relations between precipitation, temperature and dynamics of sporo-

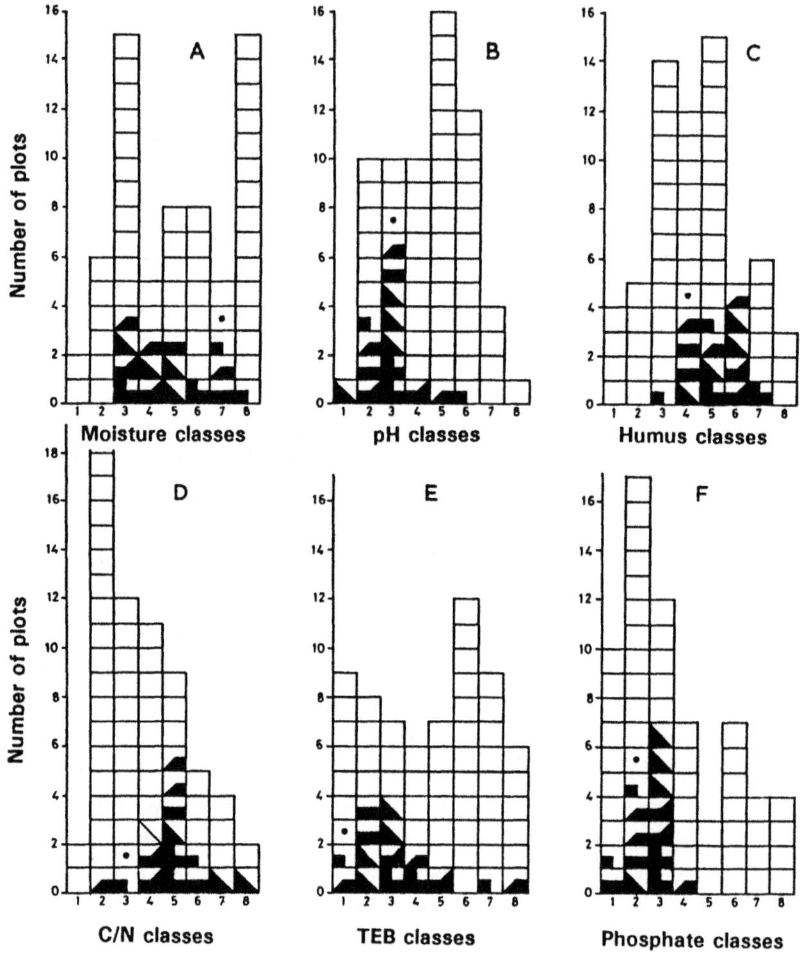

Fig. 5. Ecological spectra of *Mycena epipterygia* (Scop.: Fr.) S.F. Gray va. *epipterygia* in grasslands and moist heathlands in Drenthe, the Netherlands (after Arnolds 1982).
A = Moisture classes (1: wet . . . 8: very dry).
B = pH Classes (1: pH H_2O 3.0—3.5 . . . 8: pH H_2O 6.5—7.0).
C = Humus classes (1: organic matter content (loss-on-ignition) 0.75—1.5% . . . 8: organic matter content > 35%).
D = C/N classes (1: C/N ratio 12.5—15.0 . . . 8: C/N ratio 30.0—33.0).
E = TEB classes (1: total extractable bases 0—5 me/dm^3 . . . 8: total extractable bases 120—200 me/dm^3).
F = Phosphate classes (1: extractable phosphate (P-Olsen) 6—9 mg/dm^3 . . . 8: extractable phosphate 100—170 mg/dm^3).
Each square represents one plot. The species was only present in plots with a symbol. The symbols indicate the maximum density per visit during 3 years of study according to a logarithmic scale (small dot = <3 sporocarps per 1000 m^2 . . . lower half black 30—100 sporocarps per 1000 m^2 . . . entire square black = >10 000 sporocarps per 1000 m^2 (see Arnolds 1982: 20).

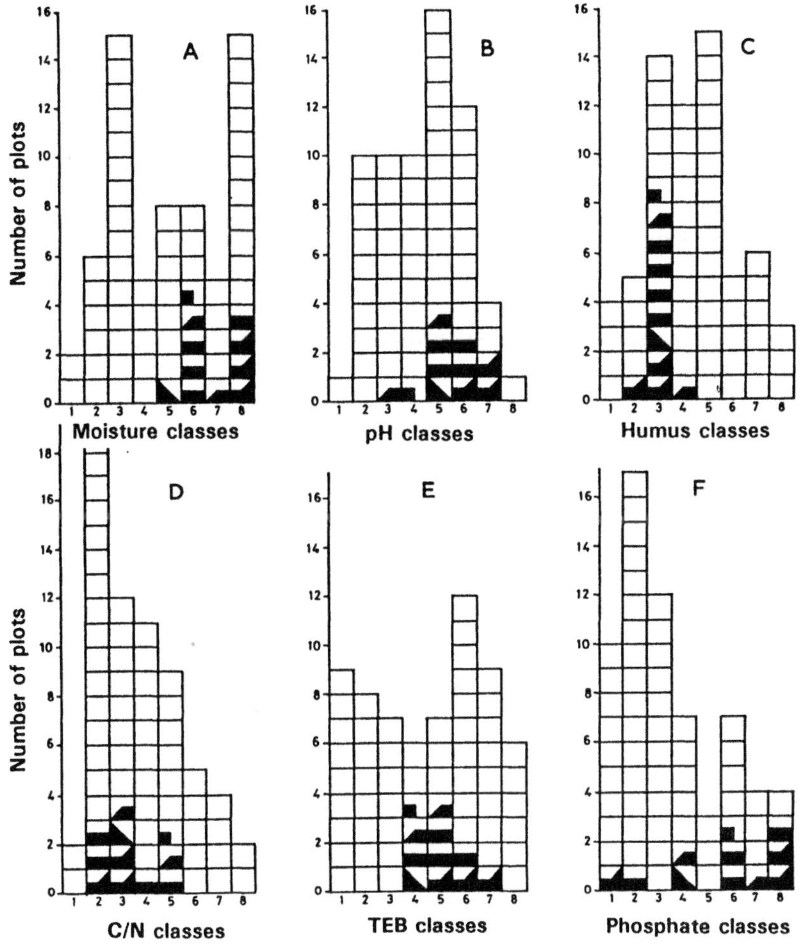

Fig. 6. Ecological spectra of *Mycena flavoalba* (Fr.) Quél. in grasslands and moist heathlands in Drenthe, the Netherlands (after Arnolds 1982).
For explanation of symbols, see Figure 5.

carp formation in grasslands and bogs, e.g. Wilkins and Patrick (1940), Lange (1948, 1984), Arnolds (1981, 1988b), Brunner (1987) and Brunner and Horak (1988). Moisture conditions in the top soil and the air near the soil surface appear to be the most important factors for fruiting. Dry periods may terminate the fruiting process completely (Lange 1984) and it takes about 7 to 10 days before sporocarps appear after heavy rainfall. In constantly humid environments fruiting is less influenced by precipitation, for instance in arctic areas (Petersen 1977) and wet parts of bogs. The fluctuations are stronger and more dependent on rainfall in open communities than in forests, not only because of the more variable microclimatological conditions, but also because lignicolous and ectomycorrhizal fungi in forests can utilize water stored in rotten wood and the mycelial mantle, respectively (Lange 1984). However, only part of the fluctuations in fruiting can be explained by the preceding short-term weather conditions (Arnolds 1981). Other factors, such as productivity of the plant community, availability of appropriate substrates, exhaustion of mycelia by abundant fruiting in the previous year, and weather conditions during the winter months, may be important as well (Arnolds 1988b).

The periodicity of fruiting of different species (called autoperiodicity by Arnolds 1981) was discussed among others by Arnolds (1982), Lange (1984) and Brunner and Horak (1988). Examples of periodicity diagrams are presented in Figure 7, based on research in grass- and heathlands in the Netherlands. *Agrocybe paludosa* (J. Lange) Kühn. & Romagn. is a vernal species (Figure 7), *A. semiorbicularis* (Bull.) Fayod is typical of summer (Figure 7), *Bovista nigrescens* Pers.: Pers. has its optimum in early autumn (Figure 7) and *Mycena cinerella* P. Karst. in late autumn (Figure 7). *Marasmius oreades* is observed from May until late November, but least in summer (Figure 7). *Psilocybe muscorum* P. D. Orton is fruiting during the entire autumn and has a secondary, lower peak in spring (Figure 7).

5.5.3. *Human influence*

Macrofungal communities outside forests are often strongly influenced by human activities, with different results. Disturbance of critical environmental factors which causes the destruction of a certain plant community, necessarily leads to drastic changes in the accompanying macrofungocoenosis. Examples of such interference are the drainage of bogs and wetlands, the reclamation and afforestation of heathlands and irrigation of dry steppes.

On the other hand, most grass- and heathlands in Northwest and Central Europe are dependent on some forms of human influence, mainly grazing, cutting and burning or a combination of these measures. The effects of grazing and cutting on the composition and structure of grassland communities have recently been reviewed by Bakker (1989). Studies concerning their influence on macrofungocoenoses are scarce and the results are in part contradictory.

Brunner (1987) compared 6 regimes of cutting in limestone grasslands (Mesobromion) in northern Switzerland. The numbers of macrofungal species decreased in the following order: plots cut every fifth year in June

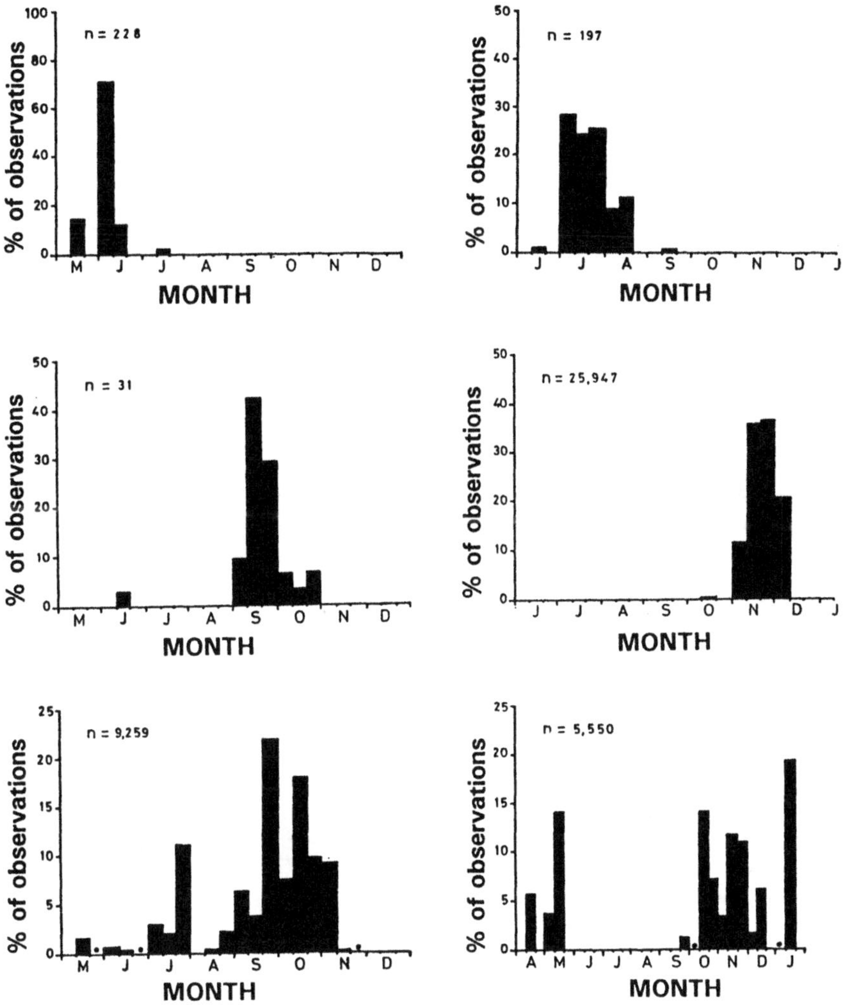

Fig. 7. Periodicity of various species of macrofungi in grass- and heathlands in the Netherlands, based on added frequency in the period 1974—1977 in 64 permanent plots (after Arnolds 1982).
n = total number of observations of a species.
A = *Agrocybe paludosa* (J. Lange) Kühn. & Romagn. — B = *Agrocybe semiobicularis* (Bull.) Fayod s. lat. — C = *Bovista nigrescens* Pers. — D = *Mycena cinerella* P. Karst. — E. = *Marasmius oreades* (Bolt.: Fr.) Fr. — F = *Psilocybe muscorum* P.D. Orton.

(average 5.0 species/50 m^2, n = 8), unmanaged plots (av. 3.9/50 m^2, n = 14), plots burnt every year in spring (av. 3.6/50 m^2, n = 7), plots cut every year in June or July (av. 2.5/50 m^2, n = 17), plots cut every year in October (av. 2.4/50 m^2, n= 8) and plots cut every second or third year in June–September (av. 2.1/50 m^2, n = 20). The productivity of sporocarps was highest in unmanaged plots and lowest in the plots cut every year in October or every second or third year. On the other hand, Sadowska (1973) found a considerably higher number of species (18) and sporocarps (683) in a mown plot (50 m^2) of Arrhenatheretum on clay in Poland, than in an adjacent unmown plot (6 species and 12 sporocarps). Also Arnolds (1980) observed a strong reduction of species diversity and fruiting in unmanaged, semi-natural grasslands compared with grazed or cut grasslands. Kuyper and Schreurs (1984, Table III) found a very rich mycoflora in a limestone grassland in the Netherlands few years after resumption the traditional practice of sheep grazing. In the former, unmanaged grassland, a few species were found in low densities. In this case the mycelia were probably still present in the soil and only fruiting seems to be inhibited by a dense and tall sward and/or accumulation of litter. Nitare (1988) mentioned the abandoning of pastures and afterwards spontaneous wood regeneration as one of the main factors for the disappearance of valuable grassland macrofungocoenoses in Sweden. Arnolds (1981) found a negative correlation between the productivity of sporocarps and the height of the sward in the Netherlands. Well-developed macrofungocoenoses are known from frequently mown lawns (Bond 1981). Fruiting in some species seems to be stimulated by mowing. This was noticed by Kalamees (1981) for *Panaeolina foenesecii* (Pers.: Fr.) Maire and *Panaeolus subbalteatus* (B. & Br.) Sacc. in hayfields in Estonia.

Consequently, it seems that in most grasslands a short and open structure of the sward, obtained by grazing or cutting, is favourable for most macrofungi. The reverse may be true for xerothermic grasslands on very poor soils, where a tall sward may create favourable microclimatological conditions (possibly the case in the plots, studied by Brunner 1987). In addition a positive relation exists between the development of the moss layer and the species diversity of macrofungi (Arnolds 1981, Senn-Irlet 1987). A well-developed moss layer is also promoted by mowing or grazing.

The influence of fertilization on grassland fungi was experimentally studied by Lange (1982). He compared the effects of $Ca(NO_3)_2$, P_2O_5 and K_2O in different combinations in plots with recently sawn monocultures of grasses on a moraine clay, frequently mown according to a lawn regime. Phosphorus and potassium application did not show significant effects, but nitrogen fertilization had a significant positive effect on species diversity: on the average 9.7 species in treated plots against 4.0 species in untreated plots. Nitrogen also influenced the species composition: Lange (1982) regarded 24 species as nitrophilous, 10 as nitrophobous and 6 as indifferent. Bond (1972) described the relations between macrofungi and different levels of NPK fertilizers in an orchard with grassy undergrowth. He did not find significant differences in species diversity and few differences in composition. On the contrary, many authors described strong, deleterious effects of fertilizer application on species diversity of macrofungi in permanent pastures on former not or weakly fertilized soils (e.g. Arnolds 1980, 1981, 1989, Brunner 1987, Nitare 1988). The impoverishment of mycoflora is generally ascribed to increase of available nitrogen, but a good evaluation of the influence of other nutrients, in particular phosphorus, is impossible by lack of experimental research.

An important factor for the development of a species-rich mycoflora is the permanent, stable use of grasslands, as stressed by Neuhoff (1949, 1950), Becker (1956), Arnolds (1980) and Nitare (1988). Lange (1982) found a slightly higher number of species in 7 year old, sown grasslands than in 2 year old plots. A large number of genera and species is characteristic of many types of old, unfertilized grasslands, throughout Europe, for instance *Hygrocybe* (incl. *Camarophyllus*), *Hygrotrama*, *Entoloma* subgenus *Leptonia*, *Dermoloma*, *Clavaria*, *Clavulinopsis* and *Geoglossaceae* (Arnolds 1980, Rald 1985, Nitare 1988). It is striking that such species are completely absent from the young experimental plots, described by Lange (1982) and Bond (1972). Many rare species of the genera mentioned are restricted to grassland communities that have been undisturbed during several decades (Becker 1956, Nitare 1988); they are consequently important bioindicators from a view-point of nature conservation. Nitare (1988) proposed a scale for the evaluation of macrofungocoenoses in Swedish grasslands, based on species diversity of indicator genera (Figure 8). This scheme may be very well suitable for application in other parts of Europe.

5.6. Macrofungocoenoses in various plant communities

5.6.1. *Introduction*

In the next sections some general characteristics are given on development and composition of macrofungocoenoses in various plant communities outside forests. The state of knowledge is indicated and references to the most relevant literature are given. Surveys of characteristic fungi for plant communities in larger areas were published by Dennis (1955) for Scotland, Heinemann and Darimont (1956) for Belgium, Bon and Géhu (1973) for France, Kalamees (1978, 1980) for the Estonian S.S.R., Kreisel and Dörfelt (1985) for the German Federal Republic, and Arnolds (1988c) for the

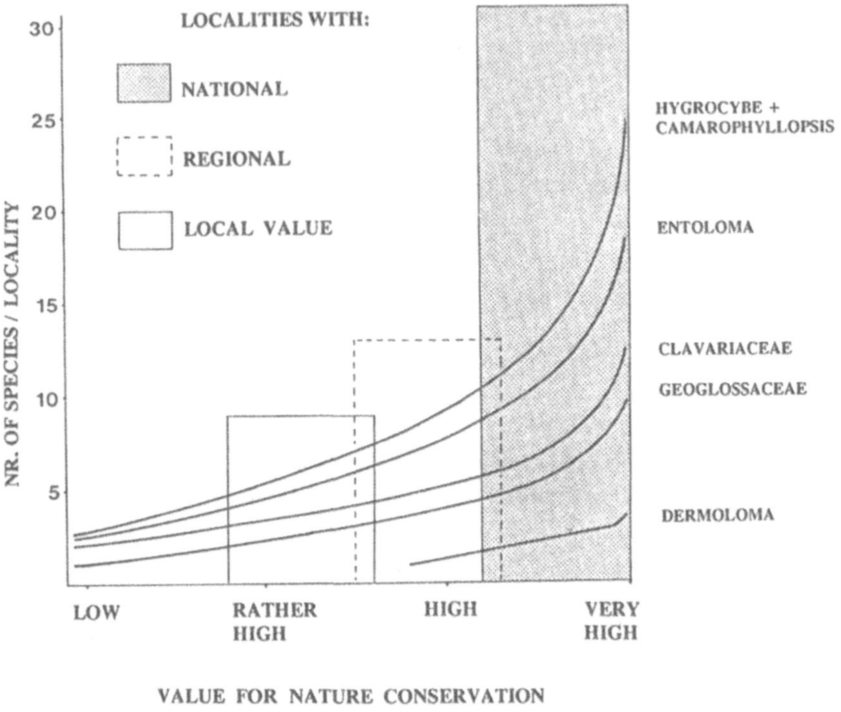

Fig. 8. A scheme for assigning the value for nature conservation to seminatural grasslands in Sweden according to the number of species of grassland fungi (after Nitare 1988).

Netherlands. These surveys are based on mycofloristic and/or mycocoenological data and vary in accuracy. They do not contain quantitative data on the presence or abundance of species.

5.6.2. *Macrofungocoenoses in grasslands*

The first mycocoenological studies in grasslands were carried out by Wilkins and Patrick (1939, 1940) in England. Although their studies can be assigned to the classics in mycocoenology, the results are of limited value because of strongly changed species concepts.

In more recent years, extensive comparative studies in various grassland-types were published by Winterhoff (1975), Guminska (1976), Arnolds (1981, 1982) and Brunner (1987). In addition, a number of more local and mycofloristical studies appeared (Table III). However, most grassland types were investigated only once and in a restricted area. The available data do not cover more than a fraction of the variation in European grassland communities. Dry, poor grasslands on calcareous loam and sand have been studied most thoroughly. Few exact data are available on wet grasslands and no quantitative figures were published from grasslands in coastal dunes, halophyte communities and natural steppe communities in eastern Europe. Mycofloristic-ecological studies were made by e.g. Bon (1970, coastal communities in northern France), Courtecuisse (1984, 1986, idem), Einhellinger (1973, pastures in southern Germany), Kreisel (1957, various grasslands in the coastal G.D.R.) and Moravec (1960, serpentine steppe in Czechoslovakia).

The species diversity and sporocarp abundance of macrofungi are comparatively low in pioneer grasslands on sand dunes (Corynephorion) and wet hayfields (Molinietalia), high in mesic to dry, unfertilized communities on weakly acid to basic soils (Table III). The most xerophytic communities are characterized by a high proportion of gasteromycetes, which are almost lacking in wet hayfields. *Entoloma* and *Hygrocybe* are dominant genera in moss-rich meadows on various types of mesic to dry, unfertilized soils. *Conocybe* species prefer weakly acid to basic and fertilized grasslands; *Galerina*, on the other hand, acid, unfertilized communities. In strongly fertilized grasslands most genera are poorly represented, but *Conocybe* and *Panaeolus* are relatively important.

The relations between macrofungocoenoses and different types of grassland exploitation were discussed in § 5.5.3.

5.6.3. *Macrofungocoenoses in heathlands*

Thorough mycocoenological studies in moist to dry heathlands were carried out in the Netherlands (Kramer 1969, Arnolds 1981, 1982, 1988b), the German Federal Republic (Runge 1960) and Poland (Rudnicka-Jezierska 1969). Heathlands are characterized by a uniform litter layer, colonized by a low to moderate number of macrofungal species, but some of them (mainly *Mycena* spp. and *Marasmius androsaceus*) fruiting in high densities. *Mycena*, *Galerina* and *Entoloma* are the genera represented with most species. Grass heaths are very similar, but slightly richer in species (Table IV) and in composition more related to grasslands (Arnolds 1981). On the other hand, some litter-inhabiting species (e.g. *Entoloma helodes* (Fr.: Fr.) Kummer, *Hypholoma elongatipes* Peck, *H. udum* (Pers.: Fr.) Kühn.) are shared with bog communities, although some authors regarded them as strictly sphagnicolous.

5.6.4. *Macrofungocoenoses in marshes and bogs*

Aquatic ecosystems are virtually without macrofungi except the rare *Cordyceps entomorrhiza* (Dickson) Link, a parasite on beetle larvae (Petch 1948).

Various types of marshland accommodate characteristic macrofungocoenoses, but quantitative data on their composition and the abundance of sporocarps are scarce. Favre (1948) and Einhellinger (1976, 1977) carried out extensive mycofloristic-ecological investigations in bogs in the Swiss Jura and Bavaria, respectively. They published lists of species, considered as characteristic of (groups of related) plant communities on the basis of annotated foray lists from a number of bogs with foray frequency as quantitative parameter. Favre (1948) mentioned, for example, 22 species of macrofungi as exclusive for 'Sphagnetum' and in addition a larger number of accidental species; 30 species as occurring in the Magnocaricion and 70 for the *Filipendula ulmaria* association. Einhellinger (1976) listed for 'Sphagnetum' 19 characteristics and 30 accidental species, for the Magnocaricion 17 and for the Caricion davallianae 72 species. These data suggest a well-developed mycoflora, but in reality many species are rare and sporocarp density is low in many areas. This is shown in the more exact, although incomplete data for bog communities in Austria by Krisai (1987) (Table IV). Similar results were obtained during more intensive research by Barkman (in prep.) in the Netherlands. Augustin and Runge (1968) found a low species diversity, but a high sporocarp density in a bog community in northern Germany. Lange (1948) made an extensive analyses of the mycocoenoses in a bog in Denmark, by means of plots of the unusual size of 1 m^2. He found amazingly high numbers of species and sporocarps (Table IV). Kreisel (1954) described the macrofungal flora of bogs in Mecklenburg (G.D.R.).

The macrofungal flora in oligotrophic peat bogs is dominated by the agaric genera *Galerina*, *Mycena*, *Hypholoma* and *Omphalina*. For some species of *Tephrocybe* and *Galerina* a parasitic relationship with *Sphagnum* has been demonstrated (§ 5.3.4). Species of *Hypholoma* also grow on bare peat and heath litter and are presumably saprophytes on substrates with a high C/N ratio, including dead *Sphagnum*. Some species of the, usually ectomycorrhizal, genus *Cortinarius* (subgenus *Dermocybe*) have been reported from both *Sphagnum*-rich forests and *Sphagnum* communities without any ectomycorrhizal trees (Favre 1948, Einhellinger 1976, Høiland 1983). They are regarded as facultative ectomycorrhizal symbionts, but their actual life strategy outside forests and scrubs is unknown.

In more eutrophic marshes species of e.g. *Mycena*, *Marasmius* and *Entoloma* are more important, together with small discomycetes of the

Helotiales. They live as saprophytes on fresh and decaying litter of herbaceous plants.

5.6.5. *Macrofungocoenoses in alpine and arctic plant communities*

Favre (1955) was the first author to draw attention to the rich, diverse and specialized macrofungal flora above the timberline in his classical myco-floristic-ecological study in the Swiss National Park. He characterized some dwarf shrub communities as 'microsilvae' in view of their mycofloristic affinities with forest communities and the importance of ectomycorrhizal associations. Enumerations of species in different alpine communities were given without quantitative data.

Lange (1957) found similar results when analyzing plots in arctic dwarf shrub communities in Greenland. His quantitative figures, based on incidental inventories of plots of 1 m^2, are difficult to compare with most other mycocoenological data (Table IV). More complete plot studies were made by Senn-Irlet (1987) in the Swiss Alps and by Metsänheimo (1987) in arctic Finland. In addition a number of autecological data were published, also by taxonomists (e.g. Petersen 1977, Laursen and Ammirati 1982, Laursen *et al.* 1987, Gulden *et al.* 1985).

In particular rich in species and sporocarps are seepage communities with *Salix* spp., snowbed communities with dwarf *Salix* on calcareous soils and some types of *Betula nana* heath (Table IV). Ectomycorrhizal species make up about 40 to 65% of the total number of species and even a higher proportion of the annual productivity. Snowbed communities on acid soil are somewhat poorer in species and sporocarps. The ectomycorrhizal species are there equally important. Dry alpine grasslands on basic soil (e.g. Seslerio-Caricetum sempervirentis) are also rather poor in species. Ectomycorrhizal species are less frequent and mainly associated with *Dryas*. Alpine grasslands on dry, acid soil are remarkably poor in macrofungi and without mycorrhizal species. Ectomycorrhizal associations between *Salix* spp. and various macrofungi were demonstrated in vitro by e.g. Antibus *et al.* (1981), between *Dryas octopetala* and two *Hebeloma* species by Debaud (1987).

5.6.6. *Macrofungocoenoses in ephemeral plant communities*

Exact data on the macrofungocoenoses in ephemeral plant communities, such as arable fields, weed and ruderal communities and flood marks are nearly lacking. Lists of 'characteristic species' by Bon and Géhu (1973) are of limited importance since they are based on accidental observations and even include for instance a number of ectomycorrhizal species, associated with trees. The most extensive research in arable fields was carried out by Hille (1983) in the German Federal Republic (Table III). According to that author the density of sporocarps is extremely low. The number of species is very low in intensively cultivated, small-scaled fields, slightly higher in perennial crops such as *Medicago sativa* and in less intensively cultivated, small-scaled fields, although potential organic substrate is available in all types (straw and dung remains). Apparently the periodic fragmentation of mycelia is destructive for most macrofungi and in fertile (specially N-rich) conditions microfungi and bacteria are better competitors. Differences in species diversity might be also due to applications of stable dung or artificial fertilizers, since many species belong to the subcoprophytic fungi. Dominant genera are *Conocybe* and *Psathyrella* (Table III). Wojewoda (1975) listed some species from less fertile fields in Poland (Table III); Courtecuisse (1986) observed 5 species of agarics on coastal flood-marks (Cakiletea).

Acknowledgements

Many thanks are due to Dr W. Gams (Baarn) and Dr Th. W. Kuyper (Wijster) for their critical reading and linguistic improvements of the manuscript.

References

Andersson, O. 1950. Larger fungi on sandy grass heaths and sand dunes in Scandinavia. Bot. Notiser Suppl. 2: 1—89.

Antibus, R. K., J. G. Croxdale, O. K. Miller, and A. E. Linkins. 1981. Ectomycorrhizal fungi of *Salix rotundifolia* III. Resynthesized mycorrhizal complexes and their surface phosphatase activities. Can. J. Bot. 59: 2458—2465.

Apinis, A. E. 1958. Distribution of microfungi in soil profiles of certain alluvial grasslands. p. 83—90. *In*: R. Tüxen (ed.) Angewandte Pflanzensoziologie. Ber. Intern. Symposium Pflanzensoziologie — Bodenkunde 1956, Stolzenau/Weser. J. Cramer, Weinheim.

Apinis, A. E. 1970. Das Verhalten der Pilze in bestimmten Grassland-Gesellschaften. p. 172—186. *In*: R. Tüxen (ed.) Gesellschaftsmorphologie. Ber. Intern. Symposium. Rinteln. Dr. W. Junk N. V., Den Haag.

Arnolds, E. 1980. De oecologie en sociologie van Wasplaten. Natura 77: 17—44.

Arnolds, E. 1981. Ecology and coenology of macrofungi in grasslands and moist heathlands in Drenthe, the Netherlands. Vol. 1. Biblthca mycol. 83. J. Cramer, Vaduz.

Arnolds, E. 1982. Ecology and coenology of macrofungi in grasslands and moist heathlands in Drenthe, the Netherlands. Vol. 2. Biblthca mycol. 90. J. Cramer, Vaduz.

Arnolds, E. 1988a. Status and classification of fungal communities. p. 153—165. *In*: J. J.

Barkman, and K. V. Sykora (eds.) Dependent Plant Communities. SPB Academic Publishing, The Hague.
Arnolds, E. 1988b. Dynamics of macrofungi in two moist heathlands in Drenthe, the Netherlands. Acta Bot. Neerl. 37: 291—305.
Arnolds, E. 1988c. The Netherlands as an environment for agarics and boleti. p. 6—29. *In*: C. Bas, Th. W. Kuyper, M. E. Noordeloos, and E. C. Vellinga (eds.) Flora Agaricina Neerlandica 1. A. A. Balkema, Rotterdam, Brookfield.
Arnolds, E. 1989. The influence of increased fertilization on the macrofungi of a sheep meadow in Drenthe, the Netherlands. Opera Botanica 100: 7—21.
Arnolds, E. and B. de Vries. 1989. Oecologische statistiek van de Nederlandse macrofungi. Coolia 32: 76—86.
Augustin, A., and A. Runge. 1968. Pilze in Scheiden-Wollgras-Rasen des Emsdettener Venns. Natur und Heimat 28: 152—153.
Bakker, J. P. 1989. Nature management by grazing and cutting. Kluwer Academic Publ., Dordrecht, Boston, London.
Barkman, J. J. 1976. Terrestrische fungi in jeneverbesstruwelen. Coolia 19: 94—110.
Barkman, J. J., and Ph. Stoutjesdijk. 1987. Microklimaat, vegetatie en fauna. Pudoc Wageningen.
Becker, G. 1956. Observations sur l'écologie des champignons supérieurs. Ann. sci. Univ. Besancon (sér. 2, Bot.) 7: 15—128.
Bell, M. K. 1974. Decomposition of herbaceous litter. p. 37—67. *In*: C. H. Dickinson, and G. J. F. Pugh (eds.) Biology of plant litter decomposition, Vol. 1. Academic Press, London, New York.
Benkert, D. 1976. Bemerkenswerte Ascomyceten der DDR I. Zu einigen Arten der Gattung *Lamprospora* De Not. Feddes Repert. 87: 611—642.
Bohus, G. 1957. On the results of researches concerning the temperature claims of macroscopic fungi. Ann. Hist. Nat. Musei nat. Hungarici 8: 79—86.
Bohus, G. 1984. Studies on pH requirement of soil-inhabiting mushrooms: the R-spectra of mushroom assemblages in deciduous forest communities. Ann. bot. Hung. 30: 155—171.
Bon, M. 1970. Flore héliophile des Macromycètes de la zone maritime picarde. Bull. trim. Soc. mycol. Fr. 86: 79—213.
Bon, M., and J. M. Géhu. 1973. Unités superieures de végetation et récoltes mycologiques. Docum. mycol. 2 (6): 1—40.
Bond, T. E. T. 1972. Observations on the macro-fungi of an apple orchard in relation to cover crops and NPK fertilizers. Trans. Br. mycol. Soc. 58: 403—416.
Bond, T. E. T. 1981. Macro-fungi on a garden lawn, 1971—78. Bull. Br. mycol. Soc. 15: 99—138.
Brunner, I. 1987. Pilzökologische Untersuchungen in Wiesen und Brachland in der Nordschweiz (Schaffhauser Jura). Veröff. geobot. Institut Eidg. Techn. Hochschule Zürich 92: 1—241.
Brunner, I., and E. Horak. 1988. Zür Ökologie und Dynamik praticoler Agaricales in Mesobrometen der Nordschweiz. Mycologia Helvetica 3: 1—26.
Courtecuisse, R. 1984. Transect mycologique dunaire sur la Côte d'Opale (France) (1ére partie: Les groupements héliophiles et arbustifs de la xerosère). Docum. mycol. 15: 1—115.
Courtecuisse, R. 1986. Transect mycologique dunaire sur la Côte d'Opale (France). II. Les groupements de l'hygrosère. Docum. mycol. 17: 1—70.
Debaud, J. C. 1987. Ecophysiological studies on alpine macromycetes: saprophytic *Clitocybe* and mycorrhizal *Hebeloma* associated with *Dryas octopetala*. p. 47—60. *In*: G. A. Laursen, J. F. Ammirati, and S. A. Redhead (eds.) Arctic and alpine mycology II. Plenum Press, New York, London.
Dennis, R. W. G. 1955. The larger fungi in the North-West Highlands of Scotland. Kew Bull. 10: 111—126.
Döbbeler, P. 1979. Untersuchungen an moosparasitischen Pezizales aus der Verwandtschaft von *Octospora*. Nova Hedwigia 31: 817—864.

Einhellinger, A. 1969. Die Pilze der Garchinger Heide. Ein Beitrag zur Mykosoziologie der Trockenrasen. Ber. Bayer. bot. Ges. 41: 79—130.

Einhellinger, A. 1973. Die Pilze der Pflanzengesellschaften des Auwaldgebiets der Isar zwischen München und Grüneck. Ber. Bayer. bot. Ges. 44: 5—100.

Einhellinger, A. 1976. Die Pilze in primären und sekundären Pflanzengesellschaften oberbayerischer Moore. Teil I. Ber. Bayer. bot. Ges. 47: 75—149.

Einhellinger, A. 1977. Die Pilze in primären und sekundären Pflanzengesellschaften oberbayerischer Moore. Teil II. Ber. Bayer. bot. Ges. 48: 61—146.

Favre, J. 1948. Les associations fongiques des hauts-marais jurassiens et de quelques régions voisines. Matér. Flore cryptogam. Suisse 10 (3). Bern.

Favre, J. 1955. Les champignons supérieurs de la zone alpine du Parc National Suisse. Ergebn. wiss. Unters. Schweiz. Nat. Parks (N.F.) 5: 1—212. Liestal.

Fellner, R., and J. Biber. 1989. *Helianthemum* and some Agaricales: unusual case of ectomycorrhizal symbiosis. Agricult. Ecosyst. Environm. 28: 121—125.

Fontana, A. 1977. Ectomycorrhizae in *Polygonum viviparum* L. Abstracts 3rd NACOM. Athens, Georgia.

Fontana, A., and G. Giovanetti. 1979. Simbiosi micorrizica fra *Cistus incanus* L. ssp. *incanus* e *Tuber melanosporum* Vitt. Allionia 23: 5—11.

Frankland, J. C., J. N. Hedger, and M. J. Swift (eds.). 1982. Decomposer basidiomycetes: their biology and ecology. Cambridge Univ. Press, Cambridge.

Fries, N. 1949. Culture studies in the genus *Mycena*. Svensk bot. Tidskr. 43: 316—342.

Gimingham, C. H. 1972. Ecology of heathlands. Chapman and Hall, London.

Glowinski, H. 1984. Zur Pilzflora des Naturschutzgebiets 'Dummersdorfer Ufer' bei Lübeck. Beitr. Kenntnis Pilze Mitteleuropas 1: 119—132.

Grunda, B. 1976. Effects of fungal 'fairy rings' on soil properties. Ceská Mykol. 30: 27—32.

Gulden, G., K. M. Jenssen, and J. Stordal. 1985. Arctic and alpine fungi, Vol. 1. Soppkonsulenten, Oslo.

Guminska, B. 1976. Macromycetes of meadows in Pieniny National Park. Acta mycol. 12: 3—75.

Harley, J. L., and S. E. Smith. 1983. Mycorrhizal symbiosis. Academic Press, London, New York.

Heikkilä, H., and P. Kallio. 1966. On the problems of subarctic basidio-lichens, I. Ann. Univ. Turku, Ser. A, II. Biol.-Geograph. 36: 48—74.

Heikkilä, H., and P. Kallio. 1969. On the problems of subarctic basidio-lichens, II. Ann. Univ. Turku, Ser. A, II. Biol.-Geograph. 40: 90—97.

Heinemann, P. 1956. Les landes à *Calluna* du district picardo-barbançon de Belgique. Vegetatio 7: 99—147.

Heinemann, P., and F. Darimont. 1956. Premières indications sur les relations entre les groupements végétaux et les champignons en Belgique. Nat. belg. 37: 141—155.

Hille, M. 1983. Untersuchungen über die Makromyzeten-Flora von Ackerstandorten im Gebiet der Querfurter Platte. Hercynia N.F. 20: 219—258.

Hintikka, V. 1960. Das Verhalten einiger *Mycena*-Arten zum pH sowie deren Einfluss auf die Azidität der Humusschicht der Wälder. Karstenia 5: 107—121.

Horak, E. 1985. Die Pilzflora (Makromyceten) und ihre Ökologie in fünf Pflanzengesellschaften der montan-subalpinen Stufe des Unterengadins (Schweiz). Ergebn. wiss. Unters. Schweiz. Nat. Parks (N.F.) 12: 337—467. Lüdin AG, Liestal.

Hudson, H. J. 1968. The ecology of fungi of plant remains above the soil. New Phytologist 67: 837—874.

Høiland, K. 1983. *Cortinarius* subgenus *Dermocybe*. Opera Botanica 71: 5—113.

Høiland, K., and R. Elven. 1980. Classification of fungal synedria on coastal sand dunes at Lista, South Norway, by divisive information analysis. Norw. J. Bot. 27: 23—29.

Kalamees, K. 1978. Eesti niitude seenkond ja selle sesoonne dünaamika (The fungal cover and its seasonal dynamics of the Estonian meadows). Eesti Looduseuurijate seltsi aastaraamat 67: 38—54.

Kalamees, K. 1980. The composition and seasonal dynamics of the fungal cover on mineral soils. Scripta mycol. 9: 5—70.

Kalamees, K. 1981. Fungal cover of anthropogenous sites in Estonia. p. 95—107. *In*: Anon. (eds.) Anthropogenous changes in the plant cover of Estonia. Academy of Sciences of the Estonian S.S.R., Tartu.

Kost, G. 1984. Moosbewohnende Basidiomyceten I. Morphologie, Anatomie und Oekologie von Arten der Gattung *Rickenella* Raithelh.: *Rickenella fibula* (Bull.: Fr.) Raithelh., *R. aulacomniophila* nov. spec., *R. swartzii* (Fr.: Fr.) Kuyp. Z. Mykol. 50: 215—240.

Kramer, R. N. A. 1969. Inventarisatie van *Calluna vulgaris*- en *Empetrum nigrum*-heiden op Fungi. Unpubl. report Biol. Station Wijster.

Kreisel, H. 1954. Beobachtungen über die Pilzflora einiger Hoch- und Zwischenmoore Ost-Mecklenburgs. Wiss. Z. Univ. Greifswald, Math.-naturw. R. 3: 291—300.

Kreisel, H. 1957. Die Pilzflora des Darss und ihre Stellung in der Gesamtvegetation. Beih. Rep. Spec. Nov. Veg. 137. Beitr. Vegetationskunde 2: 110—183.

Kreisel, H. 1987. Pilzflora der Deutschen Demokratischen Republik. Basidiomycetes. Gustav Fischer, Jena.

Kreisel, H., and H. Dörfelt. 1985. Pilzsoziologie. p. 67—95. *In*: E. Michael, B. Hennig, and H. Kreisel (eds.) Handb. Pilzfr. 4 (3. Aufl.) Gustav Fischer, Jena.

Kreisel, H., and G. Ritter. 1985. Ökologie der Grosspilze. p. 9—47. *In*: Michael, Hennig, and Kreisel (eds.) Handb. Pilzfr. 4 (3. Aufl.) Gustav Fischer, Jena.

Krisai, I. 1987. Ueber den sommerlichen Pilzaspekt in einigen subalpinen Mooren des Oberen Murtales (hauptsächlich des östl. Lungaus) (Österreich). Nova Hedwigia 45: 1—39.

Kuyper, Th. W., and J. Schreurs. 1984. Enkele opmerkingen over de paddestoelenflora van de Bemelerberg. Publ. Natuurhist. Gen. Limburg 34: 53—55.

Lange, M. 1948. The agarics of Maglemose. Dansk bot. Ark. 13: 1—141.

Lange, M. 1957. Denn botaniske ekspedition til Vestgrønland 1946. Macromycetes Part 3. I. Greenland Agaricales (pars), macromycetes caeteri. II. Ecological and plant geographical studies. Medd. Grønl. 148: 1—125.

Lange, M. 1982. Fleshy fungi in grass fields. Dependence on fertilization, grass species, and age of field. Nord. J. Bot. 2: 131—143.

Lange, M. 1984. Fleshy fungi in grass fields. II. Precipitation and fructification. Nord. J. Bot. 4: 491—501.

Largent, D. L., N. Sugihara, and G. Wishner. 1980. Occurrence of mycorrhizae on ericaceous and pyrolaceous plants in Northern California. Can. J. Bot. 58: 2275—2279.

Laursen, G. A., and J. F. Ammirati (eds.). 1982. Arctic and alpine mycology: The first international symposium on arcto-alpine mycology. University of Washington Press, Seattle, London.

Laursen, G. A., J. F. Ammirati, and S. A. Redhead (eds.). 1987. Arctic and alpine mycology. II. Plenum Press, New York, London.

Lindeberg, G. 1944. Über die Physiologie ligninabbauender Bodenhymenomyzeten. Symb. bot. Ups. 8: 1—183.

Magan, N. 1988. Patterns of fungal colonisation of cereal straw in soil. p. 119—126. *In*: L. Boddy, R. Watling, and A. J. E. Lyon (eds.) Fungi and ecological disturbance. Proc. Roy. Soc. Edinburgh, Section B, 94: 119—126.

Metsänheimo, K. 1987. Sociology and ecology of larger fungi in the subarctic zones in northwest Finnish Lapland. p. 61—70. *In*: G. A. Laursen, J. F. Ammirati, and S. A. Redhead (eds.) Arctic and alpine mycology II. Plenum Press, New York, London.

Meijden, R. van der, E. Arnolds, F. Adema, E. Weeda, and C. Plate. 1983. Standaardlijst van de Nederlandse Flora 1983. Rijksherbarium, Leiden.

Mikola, P. 1956. Studies on the decomposition of forest litter by basidiomycetes. Comm. Inst. For. Fenniae 69: 4—48.

Modess, O. 1941. Zur Kenntnis der Mykorrhizabildner von Kiefer und Fichte. Symb. bot. Upsal. 5: 1—147.

Moravec, Z. 1960. The Mohelno serpentine steppe. Ceská Mykol. 14: 101—108.

Neuhoff, W. 1949. Die Pilzflora holsteinischer Viehweiden in den Jahren 1946—48 (1). Z. Pilzk. 4: 1—16.
Neuhoff, W. 1950. Die Pilzflora holsteinischer Viehweiden in den Jahren 1946—48 (2). Z. Pilzk. 5: 8—12.
Nitare, J. 1988. Jordtungor, en svampgrupp på tillbakegång i naturliga fodermarker. Svensk bot. Tidskr. 82: 341—368.
Petch, T. 1948. A revised list of British entomogenous fungi. Trans. Br. mycol. Soc. 31: 286—304.
Petersen, P. M. 1977. Investigations on the ecology and phenology of the macromycetes in the Arctic. Medd. Grønl. 199: 1—72.
Pirk, W., and R. Tüxen. 1957. Höhere Pilze in nw.-deutschen *Calluna*-Heiden (Calluneto-Genistetum typicum). Mitt. flor.-soz. Arbeitsgem. N.F. 6/7: 127—129.
Rald, E. 1985. Vokshatte som indikatorarter for mykologisk vaerdifulde overdrevslokaliteter. Svampe 11: 1—9.
Read, D. J. 1974. *Pezizella ericae* sp. nov. the perfect state of a typical mycorrhizal endophyte of Ericaceae. Trans. Br. mycol. Soc. 65: 381—383.
Redhead, S. A. 1981. Parasitism of bryophytes by agarics. Can. J. Bot. 59: 63—67.
Redhead, S. A., and K. W. Spicer. 1981. *Discinella schimperi*, a circumpolar parasite of *Sphagnum squarrosum*, and notes on *Bryophytomyces sphagni*. Mycologia 73: 904—913.
Rudnicka-Jezierska, W. 1969. Higher fungi of the inland dunes of the Kampinos Forest near Warsaw. Monogr. bot. 30: 3—116.
Runge, A. 1960. Pilzökologische und -soziologische Untersuchungen in den Bockholter Bergen bei Münster. Abh. Landesmus. Naturk. Münster 22: 3—21.
Sadowska, B. 1973. Preliminary evaluation of the productivity of fungi (Agaricales and Gasteromycetes) on the Kazun meadows. Acta mycol. 9: 91—100.
Sadowska, B. 1974. Preliminary analysis of productivity of fruiting fungi on Strzeleckie meadows. Acta mycol. 10: 143—158.
Sanders, F. E., B. Mosse, and P. B. Tinker (eds.). 1975. Endomycorrhizas. Academic Press, London, New York, San Francisco.
Senn-Irlet, B. J. 1987. Oekologie, Soziologie und Taxonomie alpiner Makromyzeten (Agaricales, Basidiomycetes) der Schweizer Zentralalpen. Diss. Bern.
Seviour, R. J., R. R. Willing, and G. A. Chilvers. 1973. Basidiocarps associated with ericoid mycorrhizas. New Phytol. 72: 381—385.
Sewell, G. W. F. 1959a. Studies of fungi in a *Calluna*-heathland soil, I. — Vertical distribution in soil and on root-surfaces. Trans. Br. mycol. Soc. 42: 343—353.
Sewell, G. W. F. 1959b. The ecology of fungi in *Calluna*-heathland soils. New Phytologist 58: 5—15.
Stangl, J. 1970. Das Pilzwachstum in alluvialen Schotterebenen und seine Abhängigkeit von Vegetationsgesellschaften. Z. Pilzk. 36: 209—255.
Swift, M. J., O. W. Heal, and J. M. Andersson. 1979. Decomposition in terrestrial ecosystems. Studies in Ecology 5. Blackwell, Oxford, London, Edinburgh, Melbourne.
Warcup, J. H. 1951a. Studies on the growth of basidiomycetes in soil. Ann. bot. 15: 305—317.
Warcup, J. H. 1951b. The ecology of soil fungi. Trans. Br. mycol. Soc. 34: 376—399.
Warcup, J. H., and P. H. B. Talbot. 1962. Ecology and identity of mycelia isolated from soil. Trans. Br. mycol. Soc. 45: 495—518.
Watling, R. 1978. The distribution of the larger fungi in Yorkshire. Naturalist 103: 39—57.
Webster, J. 1956. Succession of fungi on decaying cocksfoot culms I. J. Ecol. 44: 517—544.
Webster, J. 1957. Succession of fungi on decaying cocksfoot culms II. J. Ecol. 45: 1—30.
Webster, J., and N. J. Dix. 1960. Succession of fungi on decaying cocksfoot culms III. Trans. Br. mycol. Soc. 43: 85—99.
Wilkins, W. H., and S. H. M. Patrick. 1939. The ecology of larger fungi III. Constancy and frequency of grassland species with special reference to soil types. Ann. appl. Biol. 26: 25—46.

Wilkins, W. H., and S. H. M. Patrick. 1940. The ecology of the larger fungi IV. The seasonal frequence of grassland fungi with special reference to the influence of environmental factors. Ann. appl. Biol. 27: 17—34.
Winterhoff, W. 1975. Die Pilzvegetation der Dünenrasen bei Sandhausen (nördl. Oberrheinebene). Beitr. naturk. Forsch. Südw. Deutschland 34: 445—462.
Winterhoff, W. 1978. Bemerkenswerte Pilze in Trockenrasen des nördlichen Oberrheingebiets (Fortsetzung). Hessische Floristische Briefe 27: 41—47.
Winterhoff, W. 1983. Die Grosspilze des Wingertsbuckels bei Schwetzingen (nordbadische Oberrheinebene). Carolinea 41: 33—44.
Winterhoff, W. 1986. Zur Pilzflora der fränkischen Gipshügel. Natur und Mensch, Jahresmitt. 1986: 81—85.
Wojewoda, W. 1975. Macromycetes Ojcowskiego Parku Narodowego II. Charackterystyka socjologiczno — ekologiczno — geograficzna. Acta mycol. 11: 163—209.

6. Macrofungi on special substrates

MARIA LISIEWSKA

Abstract

A survey is given of macrofungal niche substrate groups and macrofungal communities on wood, fire places, dung, arthropods, mosses, and fruit-bodies of other fungi. The dependence of the fungal population on the kind of the substrate and on the origin of the substrate in different species of trees, mosses, animals, or fungi is reviewed in detail. Attention is payed to fungal succession on decaying wood, on fire places and on dung, to classification of lignicolous, carbophilous and coprophilous community types, to phenology of fruiting and to mycogeographical aspects.

6.1. Introduction

As heterotrophic organisms macrofungi are dependent on living hosts or on dead organic matter. Most parasitic and mycorrhiza forming macrofungi are more or less bound to special hosts. Saprophytes are more or less specialized to specific substrates such as humus, fallen leaves, fruits or catkins, standing or fallen wood, bark of standing trees, remains of herbs or mosses, decaying fruit-bodies of other fungi, fireplaces, or excrements of animals. Thus in highly organized biocoenoses, such as forests, a great number of different niche-substrate groups (Arnolds 1988) or guilds (Barkman 1987) can be distinguished (see § 1.4.4).

Fungi of most niche-substrate groups are no subjects to competition of plants nor of fungi belonging to other niche-substrate groups. But not all niche-substrate groups are markedly delimited; for instance many fungi belong to several groups, parasitic macrofungi continue their life as sapro-

W. Winterhoff (ed.), Fungi in Vegetation Science, 151–182.

phytes after their host has died, and mycorrhiza forming fungi live together with humus saprophytes in the soil.

If the hosts or the particles of dead substrates are large enough, they are settled with more than one fungus species, so that special fungal communities are formed (see chapter 1). The composition of such fungal communities may be influenced by phytoclimatic conditions as for example Winski (1987) and De Vries & Kuyper (1988) have shown, but the basic factor is the presence of an adequate substrate. Therefore equal fungal communities can be found on corresponding substrates in different biocoena, and they can be described independent of the biocoena to which they belong.

As Arnolds has explained in 2.4 and 2.10 there are several conceptions of the status and classification of fungal communities on special substrates: fungi together with other cryptogams on special substrates are regarded as societies (synusiae) of the entire plant community (e.g. Lisiewska 1974). Fungal communities on special substrates are regarded and classificated as autonomic mycocoenoses (Pirk and Tüxen 1949, 1957, Ebert 1958, Kreisel 1957, 1961, Kreisel and Müller 1987, Jahn 1966, 1968, 1976, Runge 1980 and others), or they are regarded as mycosocieties, whereas the term mycocoenosis is reserved to the complete assemblage of fungi growing within a certain biocoenosis (Barkman 1973, 1976a, 1987, Arnolds 1981, 1988).

With exception of humus all individual hosts and substrates of macrofungi will disappear or become unsuitable in a more or less short period of time by ageing, death, or decomposition. Most fungi therefore are forced to infect new hosts or substrates frequently. They do this not only by airborne propagules but also by rain splash, animal vectors, spreading mycelia and rhizomorphs (see § 7.3.3, and Rayner and Boddy 1988: 147—159). Thus the infection of any host or substrate by a certain fungus may be dependent on the presence of a certain animal or of the occurrence of that fungus in the vicinity.

On decaying wood, on fireplaces, on dung, and on excrements certain sequences of fruit-body production by different species are to be observed. These sequences are interpreted by many authors as successions of fungal species or fungal communities due to competition or to change of the substrate by progressive decomposition. But one should not overlook that the observed successions are mere successions of fructification. Mycelia of late fruiting fungi may have been living in the substrate for a long time. Some parasites as *Heterobasidion annosum* and *Ustulina deusta* are known to fruit usually not until the host has died. The fruit-body succession of coprophilic fungi corresponds mainly to the different developmental period to the generative stage as Harper and Webster (1964) have shown.

Some authors have tried to establish systems of mycosynusiae or mycocoena (see below). But these systems are still provisional and very incomplete, as many mycosynusiae are still undescribed.

Depending on the type of substrate the following groups of fungi will be discussed: lignicolous macrofungi (saprophytes and parasites on wood), carbophilous macrofungi on fireplaces, coprophilous macrofungi on excrements, macrofungi living on arthropoda, on fruit-bodies of other fungal species, and muscicolous macrofungi.

6.2. Lignicolous macrofungi

Lignicolous fungi play an important role in nature as reducers — decomposers of wood. On account of the ecological and economic importance of wood decomposition the ecology of wood inhabiting fungi is well explored (see for instance Barkman, Jansen and De Vries 1983, Jahn 1990, Käärik 1974, Rayner and Boddy 1988, and Rypacek 1966).

Some lignicolous fungi attack living trees as parasites, usually specimens already weak by injury or age (e.g. *Ganoderma adspersum, Phellinus torulosus, P. ignarius*) or they grow as saproparasites both on diseased or dead parts of living trees and on dead wood (e.g. *Armillariella mellea, Auricularia mesenterica, Pleurotus ostreatus, Xylobolus frustulatus*), but the most lignicolous macrofungi are saprophytes on dead wood (Tortić and Karadelev 1986).

6.2.1. *Restriction of macrofungi to deciduous and coniferous wood with mycogeographical remarks*

Saprophytic lignicolous macrofungi are rarely restricted to a special tree species; but many species are restricted to either deciduous wood or to conifers. Parasitic fungi are more restricted to special host trees than typical saprophytes. Strid (1975) observed in the alder forests of North-Central Scandinavia that numerous species of macrofungi, growing normally on deciduous wood, occasionally could be found on coniferous wood, e.g. *Hyphodontia crustosa, H. sambuci, Plicatura nivea, Stereum hirsutum, Trametes zonata*, etc. On the other hand, a lot of species with conifers as their main substrate are known to grow on deciduous wood, e.g. *Hyphodontia breviseta, H. hastata, Fomitopsis pinicola, Gloeophyllum sepiarium* and others. Strid explained this phenomenon by the rich occurrence of mixed wood litter in many alder forests, and by the vigorous spread of lignicolous macrofungi from one piece of wood to another one.

The alders (*Alnus glutinosa* and *A. incana*) have a very rich and diverse fungus flora. Approximately 200 species of Aphyllophorales were recorded from alder wood, but only 2 species were restricted to alder species, namely *Stereum subtomentosum* and *Peniophora erikssonii.* No differences were observed in the birch species *Betula pubescens* and *B. verrucosa.* Many macrofungi show preference to birch wood, e.g. *Merulius tremellosus, Phlebia radiata, Cerrena unicolor, Trametes zonata, Fomes fomentarius, Lenzites betulina,* but only one species — *Piptoporus betulinus* — is restricted to birch wood. On *Sorbus aucuparia* mainly *Pycnoporus cinnabarinus* was predominantly recorded in Scandinavia (Strid l.c.), but it was also noticed on beech wood (*Fagus sylvatica*) in various beach associations (Lisiewska 1974). Exclusively on beech wood in lowland and mountain beech forests the following macrofungi were observed: *Marasmius alliaceus, Mycena crocata* and *Oudemansiella mucida.* Lisiewska (l.c.) considers these species as characteristic ones of the beech forests of the Fagion sylvaticae alliance Tx et Diem. 1936. On the other hand, the host tree is much more important for the occurrence of non-poroid lignicolous Aphyllophorales than the forest association (Tortić 1985). In oak-hornbeam forests on oak wood the following species of fungi can be met: *Phellinus robustus, Collybia fusipes, Daedalea quercina, Fistulina hepatica, Hymenochaete rubiginosa* and *Mycena inclinata.* In coniferous forests are some macrofungi known which grow most frequently on pine wood like e.g. *Sparassia crispa, Tricholomopsis rutilans, Phellinus pini, Trichaptum abietinum* and *Calocera viscosa.* On spruce wood there usually occur *Xeromphalina campanella, Gloeophyllum odoratum, Tyromyces caesius.* Among macrofungi accompanying fir, species occurring on different coniferous trees are encountered, but there are a number of fungi exclusively or almost exclusively associated with fir. And for example, on fir roots in mountain forests (Sudeten, Carpathian Mts., Świętokrzyskie Mts.) appear *Bondarzewia montana,* and *Sparassis nemecii,* and on trunks, stumps and branches grow the following macrofungi: *Phellinus hartigii, P. pouzarii, Hericium coralloides, Aleurodiscus amorphus, Hymenochaete mougotii* and *Hydropus marginellus.*

As Schmitt (1987) has demonstrated, the number of fungal species occupying wood of a certain tree species is dependent on the frequency of that tree, and on wood of indigenous trees more species of fungi are to be found than on wood of foreign trees.

Interesting are the observations carried out on railroad ties made of various kinds of timber in Poland. Wałek-Czernecka (1976) found on pine ties: *Lentinus lepideus, Armillariella mellea, Paxillus panuoides, Fibroporia destructor, Leptoporus mollis, Trametes versicolor, Trichaptum fusco-violaceum, Gloeophyllum sepiarium, Antrodia serialis, Dichomitus squalens, Gloeophyllum trabeum, Coniophora puteana, Corticium* (several species) and others; on oak ties: *Armillariella mellea, Schizophyllum commune, Daedalea quercina, Trametes versicolor, Stereum hirsutum, Hymenochaete rubiginosa*; on beech ties: *Chondrostereum purpureum.*

As regard to the frequence of fungal species on ties, obviously *Lentinus lepideus* is the most common. It is particularly dangerous since it can develop in the heartwood. *Daedalea quercina* is the main destroyer of oak ties. The other fungi mentioned above are of lesser importance. Some of them probably occur on nonimpregnated or badly impregnated ties.

The comparision of polypores species found in virgin forests and in cultivated forests (Kotlaba 1984) shows, that in the former there occur many different species while in the latter the number of species is lower although some of them may occur in large numbers. In cultivated (secondary) forests some species, e.g. *Heterobasidion annosum, Climacocystis borealis, Gloeophyllum odoratum* find good conditions for their growth, as well as *Abortiporus biennis* and *Ganoderma adspersum*, mostly colonize planted trees, these are the so-called synanthropic species (Kotlaba l.c.)

From the mycogeographical point of view, many lignicolous macrofungi occur practically everywhere, both in warmer and colder regions (e.g. *Schizopora paradoxa, Stereum hirsutum, Trametes versicolor, Ganoderma lucidum*), but a number of species are distinctly thermophilous and rather restricted to southern Europe (*Stereum insignitum, S. subpileatum* — Kotlaba 1985a,b), to the regions with mediterranean or submediterranean climate, e.g. the Adriatic coast and southern Macedonia, as *Phellinus torulosus, Pulchericium caeruleum, Lopharia spadicea, Peniophora lycii, Laeticorticium macrosporum, Inonotus tamariscis* and others (Tortić 1981, Tortić and Karadelev 1986).

On the basis of mycological observations in the tropics on Cuba Kreisel (1971) gives a list of vicarious species:

Holarctic	Neotropic
Fomes fomentarius	*F. marmoratus*
Pycnoporus cinnabarinus	*P. sanguineus*
Ganoderma applanatum	*G. australe*
Trametes versicolor	*T. pavonia*
Trametes gibbosa	*T. elegans*
Cerrena unicolor	*C. maxima*
Lenzites betulina	*L. cubensis*
Polyporus ciliatus	*P. tricholoma*
Oudemansiella mucida	*O. canarii*
Cyathus striatus	*C. limbatus*

On Cuba, the lignicolous macrofungi show a poor attachment to definite tree genera and for instance on deciduous wood grow the same species as on coniferous wood, palms and bamboos. On the other hand the combination of wood-inhabiting fungi is influenced by climatic factors like light and moisture.

6.2.2. *Succession of macrofungi on wood*

The succession of wood-inhabiting macrofungi, mainly living on stumps, was observed by various authors (Kreisel 1961, Jahn 1962, Ricek 1967, 1968, Runge 1975, 1978 and others). The following species of trees were taken under consideration: *Fagus sylvatica, Quercus robur, Alnus glutinosa, Betula verrucosa, B. pubescens, Tilia* spec., *Populus canadensis* and *Pinus sylvestris.* Investigations were carried out after the tree had been cut down and three succession phases were distinguished: the initial, the optimal and the final phase (Kreisel 1961).

The initial phase starts 7—10 months after tree felling and it lasts about 2 years. It is characterized by the occurrence of few pioneer species with a high degree of constancy. On stumps of different deciduous trees *Chondrostereum purpureum* and *Cylindrobasidium evolvens* have been observed in this phase. Furthermore, distinguishing species are on stumps of *Quercus: Stereum hirsutum*; on *Fagus* stumps: *Bispora monilioides* and *Schizophyllum* commune; on *Pinus* stumps: *Phlebiopsis gigantea, Trichaptum abietinum, Stereum sanguinolentum* and *Gloeophyllum sepiarium.*

The optimal phase begins after the second vegetative period and it lasts 2—5 years. In this phase grow on deciduous wood the following fungi: *Trametes versicolor, Lenzites betulinus, Bjerkandera adusta, Cerrena unicolor, Polyporus brumalis*; on coniferous wood: *Skeletocutis amorpha, Gymnopilus hybridus, Hypholoma capnoides, Tricholomopsis rutilans, Heterobasidion annosum, Tyromyces caesius, Gloeophylum odoratum.*

The final phase is observed on 5—15 year old stumps. It is the longest phase, since fungi grow also on almost decomposed stumps of different tree species. The following species are regarded as characteristic for the final phase on deciduous wood: *Xylaria hypoxylon, X. polymorpha, Ustulina deusta, Kuehneromyces mutabilis, Mycena galericulata*; and on coniferous wood: *Calocera viscosa, Pseudohydnum gelatinosum, Hypholoma fasciculare, Mycena viscosa* and others.

Strid (1975) described the succession of macrofungi on standing trees in alder forests. Infection occurs by spores of saprophytic fungi through bark cracks, broken branches, etc. The fastest decomposition of alder trunks takes place when they are infected by parasites which cause an intensive rot, e.g. *Inonotus radiatus, Phellinus ignarius, Formes fomentarius,* and others. As saprophytes they continue to decompose the wood after the death of the tree, and finally they cause the tree to fall. The mycelia penetrate the still corticated wood, and fruit-bodies are formed in bark cracks. The typical fungi on standing alders are: *Peniophora erikssonii, Hyphoderma radula, Plicatura nivea, Stereum subtomentosum,* and others. Many fungi, common on standing trunks, occur as initial species also on fallen trunks and branches. On more decayed and moist wood of fallen trunks the optimal stage of the succession has been observed. The fungi of highly decayed wood also frequently grow on other plant debris and they constitute the final stage of the decomposition of wood carried out by Aphyllophorales. These fungi are not restricted to special hosts.

The succession of macrofungi on wood is influenced considerably by microclimate, and particularly by the humidity of air and the substrate. In dry and insolated habitats, the initial phase is the longest with as dominating species: *Schizophyllum commune* and *Gloeophyllum sepiarium.* Humid microclimate accelerates the decomposition of wood and the succession of fungi.

6.2.3. *Lignicolous fungal communities*

As Darimont (1973), Jahn (1979), Keizer (1985), Winski (1987) and others have demonstrated, living and dead standing trunks, fallen trunks, stumps, still hanging and fallen dead branches and twigs of different degree of decay are inhabited by different fungal communities. Thus a great number of wood inhabiting mycosynusiae or mycocoenoses (better mycocoena) may be distinguished.

The mycosynusiae or mycocoenoses of wood-inhabiting fungi, similarly as phytocoena, have been arranged into syntaxonomic units in the rank of classes, orders, alliances and associations (Pirk and Tüxen 1957, Jahn 1966, 1968, 1979, Ricek 1967, 1968, Darimont 1973, and others). By Darimont associations are called 'sociomycies'. If fungal community types are taken as mycosynusiae the term 'association' ought to be replaced by 'union' (cf. § 2.10).

The listing of the hitherto described lignicolous fungal community types has been based on the division made in the work of Darimont (l.c.) modified by Kreisel (in Michael, Hennig and Kreisel 1981), Runge (1980) and Fellner (1988). The endings of latin names of syntaxa have been applied according to the principles used in phytosociology (Table I).

Table I. Syntaxonomic differentiation of lignicolous macrofungi

- STEREO-TRAMETETEA (= Armillarietea melleae)
  - Trametalia versicoloris
    - Marasmio-Dochmio-podion
      - Marasmietum ramealis
      - Marasmietum insititii
    - Aleuro-discion
      - Aleurodiscetum amorphi
    - Stereo-Schizo-phyllion
      - Pleurotetum cornucopiae
      - Bulgarietum polymorphae
      - Crepidotetum calolepidis
      - Stereo-Schizophylletum
    - Calo-cerion
      - Caloceretum viscosae
      - Tyromycetum caesii
      - Osmoporetum odorati
    - Pluteo-Pholiotion
      - Mycenetum galericulatae
      - Trametetum quercinae
      - Xylosphaeretum hypoxylonis
      - ?Xylarietum hypoxylonis
      - Trametetum gibbosae
      - Trametetum versicoloris
      - Trametetum hirsutae
      - Bisporetum antennatae
      - Myceno inclinatae-Hymenochaetetum rubiginosae
      - Pluteetum nani
      - Xeruletum longipedis
      - Trametetum flaccidae
      - Pholiotetum adiposae
    - Fometion
      - Phellinus robustus v. hyppophaës — Ph. contiguus Ass.
      - Phellinetum tremulae
      - Fometum igniarii
      - Phellino-Stereetum rugosi
  - Tremello-Peniopho-retalia
    - Tre-melion
      - Tremelletum mesentericae
    - Penio-phoro-Muci-dulion
      - Mucidetum mucidae

Class Stereo-Trametetea Darimont 1973 (= Armillarietea melleae Tx. et Pirk 1952)
Lignicolous saprophytic and parasitic fungal communities.
Ch.: *Stereum rugosum, Schizopora paradoxa, Tremella foliacea, Dacrymyces stillatus, Phellinus ferruginosus, Exidia glandulosa, Pleurotus ostreatus, Phlebia radiata,* and *Armillariella mellea.*

Order Tremello-Peniophoretalia Darimont 1973
Fungal communities on branches and twigs of tree-crowns in deciduous forests.
Ch.: *Peniophora quercina, Radulomyces molaris.*

Alliance Peniophoro-Mucidulion Darimont 1973
Fungal communities on dead wood of tree-crowns.
Ch.: *Oudemansiella mucida.*

Ass. Muciduletum mucidae Darimont 1973
In tree-crowns of deciduous forests; Belgium.
Ch.: *Oudemansiella mucida.*
Mountain variant with *Panellus serotinus.*

Alliance Tremellion Dariomont 1973
Fungal communities on twigs of shrubs and trees.
Ch: *Tremella mesenterica, T. lutescens, T. globospora, T. candida, Stereum rameale, Exidia recisa, E. thuretiana.*

Ass. Tremelletum mesentericae Darimont 1973
On twigs of deciduous trees and shrubs; Belgium.
Ch.: Species of the alliance.
Mountain variant with *Plicaturopsis crispa.*

Order Trametetalia versicoloris Darimont 1973
Fungal communities on trunks, stumps, fallen branches in deciduous and coniferous forests.
Ch.: *Trametes versicolor, Bjerkandera adusta, Schizophyllum commune, Mycena galericulata, Xylaria hapoxylon, Armillariella mellea, Panellus stypticus, Pluteus atricapillus, Kuehneromyces mutabilis, Hypholoma fasciculare* and others.

Alliance Fometion Darimont 1973
Fungal communities on standing trunks of deciduous trees.
Ch.: *Fomes fomentarius, Piptoporus betulinus, Phellinus torulosus, P. ignarius, Laetiporus sulphureus, Inonotus radiatus.*

Ass. Phellino-Stereetum rugosi Darimont 1973
On trunks of living and dead deciduous trees; Belgium.
Ch.: *Stereum rugosum, Piptoporus betulinus* and other species of the alliance.
Mountain variant with *Panellus serotinus.*

Ass. Fometum ignarii Pirk 1952
On living arborescent willows (*Salix alba, S. fragilis* etc.); West Germany.
Ch.: *Phellinus ignarius, Pleurotus salignus, Trametes suaveolens.*
Ass. Phellinetum tremulae Jahn 1966
On trunks of living asp (*Populus tremula*); Sweden, West Germany.
Ch.: *Phellinus tremulae.*
Ass. *Phellinus robustus* var. *hippophaës-Ph. contiguus*-Ass. Jahn 1965
On living sea buckthorn (*Hippophaë rhamnoides*) at the sea-coast; West Germany.
Ch.: *Phellinus hippophaecola* (= *Ph. robustus* var. *hippophaës*), *P. contiguus.*

Alliance Pluteo-Pholiotion Darimont 1973
Saprophytic fungal communities on stumps of deciduous trees.
Ch.: *Psathyrella hydrophila, Collybia fusipes, Pholiota squarrosa, Lycoperdon pyriforme, Crepidotus mollis, Mycena inclinata, Ustulina deusta, Coprinus disseminatus, C. micaceus,* and others.

Ass. Pholiotetum adiposae Darimont 1973
On beech stumps in Fagetum boreoatlanticum; Belgium.
Ch.: *Pholiota adiposa, Tyromyces lacteus, Stropharia hornemannii, Physisporinus sanguinolentus, Psathyrella cotonea.*
Ass. Trametetum flaccidae Darimont 1973
On oak stumps in Quercetum sessiliflorae medioeuropaeum; Belgium.
Ch.: *Lenzites betulina* f. *flaccida, Pholiota gummosa, Hebeloma radicosum, Pholiota aurivella,* and others.
Ass. Xeruletum longipedis Darimont 1973 (included to the alliance Xerulion Fellner 1988 nom prov. together with ass. Oudemansielletum nigro-radicatae Dörfelt 1974 in Kreisel *et al.* 1981 nom nud.)
On stumps in thermophilous and xerophilous oak forests (Querco-Carpinetum primuletosum veris, Querco-Lithospermetum); Belgium.
Ch.: *Xerula longipes.*
Ass. Pluteetum nani Darimont 1973
On stumps in Aceri-Fraxinetum, mountains; Belgium.
Ch.: *Pluteus nanus, P. phlebophorus, P. lutescens, P. godeyi, P. plautus, P. salicinus, P. semibulbosus, P. villosus, Hymenochaete rubiginosa, Pholiota unicolor, Sarcoscypha coccinea,* and others.
Ass. Myceno inclinatae-Hymenochaetetum rubiginosae Dörfelt 1977
On oak stumps in thermophilous oak forests; East Germany.
Ch.: *Mycena inclinata, Hymenochaete rubiginosa.*
This fungal community type is similar to Trametetum quercinae Ricek 1967.
Ass. Bisporetum antennatae Jahn 1968

On recent sections of beech stumps and logs; West Germany.
Ch.: *Bispora molinioides* (= *B. antennata*), *Bisporella pallescens, Cylindrobasidium evolvens, Chondrostereum purpureum, Ascocoryne sarcoides.*
Ass. Trametetum hirsutae (ad interim) Jahn 1968
On recent and hard stumps of deciduous trees in sunny places; West Germany.
Ch.: *Trametes hirsuta, Schizophyllum commune.*
Ass. Trametetum versicoloris Ricek 1967
On rather recent stumps of deciduous trees (*Fagus, Quercus, Acer, Carpinus, Ulmus, Fraxinus, Aesculus*); probably common in Central Europe.
Ch.: *Trametes versicolor, T. hirsuta, Lenzites betulina, Bjerkandera adusta, Panellus stypticus.*
Ass. Trametetum gibbosae Pirk et Tüxen 1957
On stumps and rotting trunks of beech tres; probably within the Fagus area in Europe.
Ch.: *Trametes gibbosa, Xylaria polymorpha, Ustulina deusta.*
Ass. ? Xylarietum hypoxylonis Pirk 1952
On dead wood of *Salix*; West Germany.
Ch.: *Xylaria hypoxylon, Cyathus striatus, Lycoperdon pyriforme.*
Ass. Xylosphaeretum hypoxylonis Ricek 1967
On stumps of deciduous trees in shady places; Austria.
Ch.: *Xylaria hapoxylon, X. polymorpha.*
Ass. Trametetum quercinae Ricek 1967
On old stumps of oak without bark; Austria.
Ch.: *Daedalea quercina, Hymenochaete rubiginosa.*
Ass. Mycenetum galericulatae Ricek 1967
On very rotten stumps of deciduous trees in shady places; Austria.
Ch.: *Mycena galericulata, M. haematopoda.*

Alliance Calocerion viscosae all. nova prov.
Proposed name for the alliance of saprophytic mycosocieties (= Tyromyceto-Osmoporion Fellner 1988 nom. prov.) on stumps of coniferous trees.
Ch.: *Calocera viscosa, Tyromyces caesius, T. stipticus, Tricholomopsis rutilans, Pseudohydnum gelatinosum, Xeromphalina campanella.*
Ass. Osmoporetum odorati Ricek 1967
On stumps of *Picea* and *Abies* in sunny or semi-shadowed places; Austria, Belgium, Luxemburg.
Ch.: *Gloeophyllum odoratum, G. sepiarium, Pseudohydnum gelatinosum.*
Ass. Tyromycetum caesii Ricek 1967
On stumps of conifers (*Picea, Abies*) in shady places; Austria.
Ch.: *Tyromyces caesius, Dacrymyces stillatus, Heterobasidion annosum,*

Tricholomopsis rutilans.
Ass. Caloceretum viscosae Ricek 1967
On rotting stumps of coniferous trees in fairly shady places; Austria.
Ch.: *Calocera viscosa, Hydropus marginellus, Xeromphalina campanella, Mycena luteoalcalina.*

Alliance Stereo-Schizophyllion Darimont 1973
Saprophytic fungal communities on fallen trunks and branches of deciduous trees.
Ch.: *Bulgaria inquinans, Stereum subtomentosum, S. insignitum, Laxitextum bicolor, Steccherinum ochraceum, Chlorosplenium aeruginosum, Hymenochaete tabacina.*

Ass. Stereo-Schizophylletum Darimont 1973
On fallen trunks and branches of deciduous trees; Belgium.
Ch.: species of the alliance.
Mountain variant with *Plicaturopsis crispa.*

Ass. Crepidotetum calolepidis Jahn 1966
On fallen trunks of *Populus tremula*; Sweden.
Ch.: *Crepidotus calolepis.*

Ass. Bulgarietum polymorphae Runge 1962
On fallen trunks and branches of oaks, as well as on other deciduous trees; West Germany.
Ch.: *Bulgaria inquinans.*

Pleurotetum cornucopiae Kreisel et K.-H. Müller 1987
On fallen trunks and stumps of *Ulmus laevis* and *U. minor*; East Germany.
Ch.: *Pleurotus cornucopiae, Coprinus radians, Armillariella bulbosa, Oxyporus latemarginatus, Flammulina velutipes, Auricularia mesenterica, Polyporus squamosus, Lycogala epidendrum.*
Subass. P. c. polyporetosum arcularii Kreisel et K.-H. Müller 1987
Subass. P. c. bjerkanderetosum adustae Kreisel et K.-H. Müller 1987
Subass. P. c. pluteetosum atricapilli-umbrosi Kreisel et K.-H. Müller 1987

Alliance Aleurodiscion all. nova prov. (= Aleurodiscion amorphi Fellner 1988 nom. prov.)
Proposed name for the alliance of saprophytic fungal communities on fallen trunks and branches of coniferous trees.
Ch.: Aleurodiscus amorphus.

Ass. Aleurodiscetum amorphi Jahn 1968
On recent wood of fallen branches of *Abies*; West Germany.
Ch.: *Aleurodiscus amorphus, Tremella mycophaga.*

Alliance Marasmio-Dochmiopodion Darimont 1973

Saprophytic fungal communities on fallen twigs and dead blades of grasses in deciduous forests.
Ch.: *Crepidotus* (*Dochmiopus*) *variabilis, Marasmius graminum, Psilocybe inquilina, P. crobula, Polyporus arcularius* var. *agaricoides.*

Ass.: Marasmietum insititii Darimont 1973

On fallen twigs in acidophilous forests (Quercetum sessiliflorae medio-europaeum, Fagetum boreoatlanticum); Belgium.

Ch.: *Marasmius instititius, Mycena stylobates, Melanotus phillipsii, Oxyporus ravidus.*

Ass. Marasmietum ramealis Darimont 1973

On fallen twigs in calciphilous forests (Querco-Carpinetum primuletosum veris, Querco-Lithospermetum, Aceri-Fraxinetum); Belgium.

Ch.: *Marasmiellus ramealis, M. amadelphus, Marasmius epiphyllus, M. rotula, Crepidotus sphaerosporus, C. subsphaerosporus, C. pubescens, C. terricola, C. luteolus.*

On the basis of several years' observations made in sawdust depots in Hungary, Babos (1981) described a new fungal community type. This unnamed mycocoenon was fructifying regularly and often in mass on the sawdust and wood wastes of deciduous trees (*Quercus, Fagus, Robinia, Populus*, etc.), as well as on those obtained from trees mixed with coniferous wood (*Pinus, Picea*). The characteristic fungal species belong to the Agaricales: *Pluteus patricius, P. variabilicolor, P. atricapillus, Leucoagaricus bresadolae, L. meleagris, Volvariella volvacea, Leucocoprinus cepaestipes,* and *Hohenbuehelia geogenia.* The composition of the fungal community does not depend on the tree species but first of all on the thickness and the heat of decomposition of the sawdust layer.

6.3. Carbophilous macrofungi

6.3.1. *Dependence of fireplace fungi on burnt areas*

Fireplaces represent some kind of ecological complexes characterized by the fact that after the fire there occur rapid changes in all edaphic and biotic factors.

An intensive nitrification, partial or complete sterilization of soil under the influence of fire, the chemical composition of ashes, impaired biological balance of the biotop and germination of dormant spores under the influence of heat, and the creation of new nutritive compounds or toxic substances contribute, according to different authors (Moser 1949, Petersen 1970, Butin and Kappich 1980, and others), to the invasion of some fungal species, called fireplace fungi.

Depending on the degree in which the particular species depend on the burnt areas, Moser (1949) divided the fireplace fungi observed in Austria into four groups:

(a) Anthracobionts — they are obligate fireplace fungi in which the creation of fruit-bodies in natural conditions is closely connected with the fireplaces. In the first place they comprise Discomycetes, such as *Geopyxis carbonaria, Peziza violacea, Plicaria leiocarpa, P. fuliginea, Lamprospora carbonaria, Ascobolus carbonarius, Anthracobia nitida, A. melanoma, Tricharia gilva, Pyronema omphalodes*, and some Agaricales like *Pholiota carbonaria*, and *Tephrocybe ambusta.*

(b) Anthracophilous fungi — they occur mainly on fireplaces which favour their fructification, although they can grow not only on fireplaces. From Ascomycetes they comprise: *Morchella esculenta* with its varieties and forms, as well as *Morchella elata* and *M. conica*, furthermore *Helotium lutescens, Trichophaea gregaria, Octospora leucoloma, Peziza anthracophila, Neottiella hettieri,* and from Basidiomycetes: *Clavaria mucida, Fayodia maura, Geopetalum carbonarium, Mycena galopoda* var. *nigra, Psathyrella pennata, P. gossypina* and others.

(c) Anthracoxenous fungi — they are accidentally met on fireplaces which do not influence their fructification, like e.g. *Tremella mesenterica, Trametes hirsuta, Pycnoporus cinnabarinus, Auriscalpium vulgare, Hypholoma capnoides.* These fungi grow on dead wood, and they probably had lived in the forest already before the fire.

The fungi of the second and third group can be regarded as facultative fireplace fungi.

(d) Anthracophob fungi — they do not tolerate fireplaces since they hinder the development of their fruit-bodies. They comprise mainly mycorrhizal fungi such as *Russula*- and *Lactarius*-species, and others. Moser counts to them the genus *Marasmius* as well.

A similar division of fireplace fungi observed in Denmark was done by Petersen (1970). He distinguished four groups:

Group A: Species which occur exclusively on burnt ground (*Anthracobia humillima, A. maurilabra, A. melaloma, Ascobolus carbonarius, Humaria abundans, Peziza anthracina, P. echinospora, P. petersii, P. trachycarpa, P. violacea, Pustulina rosea, Sphaerosporella hinnulea, Tricharina gilva, Geopetalum carbonarium, Pholiota carbonaria, Tephrocybe carbonaria*, and *T. murina*).

Group B: Species which occur under natural conditions exclusively on burnt ground, but they may also occur on unburnt ground if the natural conditions have been somehow disturbed (*Ascobolus pusillus, Geopyxis carbonaria, Humaria gregaria, H. hemisphaerioides, Lamprospora dictydiola, Peziza atrovinosa, P. praetervisa, P. endocarpioides*).

Group C: Species which under natural conditions are common on burnt ground, but which under certain circumstances occur on unburnt ground (*Iodophanus carneus, Rhizina inflata, Fayodia maura, Coprinus angulatus, Omphalina pyxidata, Psathyrella gossypina* and *Ripartites tricholoma*).

Group D: Species which occasionally occur on burnt ground, but which are more common on unburnt ground (*Aleuria aurantia, Ascobolus geophyllus, Cheilymenia crucipila, Helvella crispa, H. lacunosa, Morchella esculenta, Octospora humosa*, and others).

Group A and B correspond to Moser's (1949) 'Anthrakobionte Pilze' — obligate fireplace fungi, and to Ebert's (1958) 'Kohlestete Arten' — constant species. Groups C and D correspond roughly to Moser's 'Kohlestete und zufällige Arten' — frequent and occasional species (Petersen 1970).

6.3.2. *Succession of macrofungi on burns*

The succession of vegetation and fungi on fireplaces was observed by many authors (Moser 1949, Pilát 1969, Petersen 1970, Butin and Kappich 1980, Turnau 1984, and others).

Moser (1949), basing on the work of Grabherr (1936), distinguished the following stages of vegetation succession on fireplaces in Austria:

I. In the first month after fire the fireplace is devoid of vegetation. The pioneers are algae, mainly *Coccomyxa*, and some Discomycetes.

II. The second stage consists of Bryophyta — *Funaria hygrometrica* and *Marchantia polymorpha* which reach their maximum development 1—2 years after fire. Then there appear the following species of fungi: *Lamprospora carbonaria, Clavaria mucida, Lopharia spadicea* (= *Stereum spadiceum*), *Rhizina inflata, Helotium lutescens* and *Trichophaea gregaria*.

III. The third stage consists of mosses: *Bryum argenteum* and other *Bryum* species, then *Ceratodon purpureus* and *Leptobryum piriforme* which indicate a 2—3-year-old fireplace. In that period *Funaria* and *Marchantia* slowly retreat and finally completely disappear. The optimum is reached by such fungi as *Octospora leucoloma, Coprinus angulatus* and *Pholiota* (*Flammula*) *carbonaria*.

IV. The fourth stage is characterized by the appearence of the first nitrophilous plants which gradually supersede mosses. In place of anthracobionts and athracophilous fungi there occur anthracoxenous. There is a marked participation of the species of *Psathyrella* and *Conocybe*.

On old fireplaces overgrown with vascular plants there occur abundantly: *Morchella conica, Gyromitra esculenta*, and *Rhizina inflata* (Pilát 1969).

Moser (1949) called attention to the differences in the species composition of fungi of fireplaces in deciduous and coniferous forests.

Petersen (1970) divided fireplace fungi into four groups depending on the period when the fireplace was settled by them:

Group I consists of species appearing on the fireplace 7 weeks after the fire and occurring not longer than 80 weeks after fire (*Anthracobia* sp. div.).

Group II consists of species occurring from 10 to 15 weeks after fire (*Peziza* sp. div., *Ascobolus carbonarius, Geopyxis carbonaria, Rhizina inflata, Tephrocybe carbonaria, Pholiota carbonaria, Rhizina inflata, Tephrocybe carbonaria, Pholiota carbonaria, Coprinus angulatus, Psathyrella gossypina*).

Group III contains species appearing 20—50 weeks after fire (*Humaria hemisphaeroides*, and *Peziza endocarpoides*).

Group IV contains species occurring from 50 to 200 weeks after fire (*Fayodia maura, Ripartites tricholoma, Geopetalum carbonarium, Omphalina pyxidata*, and others).

The succession of fungi on fireplaces investigated in Danmark is the following (Petersen l.c.):

— 7—25 weeks after fire there occur fungi of groups I and II,
— 25—100 weeks after fire species of group II dominate,
— 100—200 weeks after fire species of group IV dominate.

Interesting are the observations made by Turnau (1984) on the successive appearance of particular organisms on fireplaces in Poland (Gorce Mts.). Several days after fire the first to appear were bacteria of the so-called zymogenous type (*Bacillus mycoides*). After the first week there appeared the conidia stage of the first cup-fungus — *Trichophaea abundans*. At the end of the first month algae appeared, towards the end of the sixth week juvenile stages of mosses were observed (*Funaria hygrometrica, Bryum argenteum, Ceratodon purpureus*). Several months after the mosses had appeared, the presence of a liverwart (*Marchantia polymorpha*) was noted. Towards the end of the second month several ubiquitous species of fungi, defined as the so-called ruderal species (of Mucorales and Sphaeriales) and first fruit-bodies of cup-fungi were observed. In the second year the fruit-bodies of Basidiomycetes (*Thelephora terrestris, Pholiota carbonaria, Tephrocybe anthracophila, Omphalina* sp.) were recorded for the first time, while in the third year vascular plants entered the burns.

Special attention was paid to the periodicity of the fructification of the particular species of cup-fungi on burns as the succession progressed. Four groups of species appearing in succession on burns were distinguished:

I. Species recorded in the first weeks after burning (*Trichophaea abundans, Anthracobia* sp. div.). These species fructify exclusively on fireplaces and they depend the most on the physico-chemical conditions of these places.

II. Species appearing in the first weeks after burning and fruiting abundantly in the first year, then less numerously (*Geopyxis carbonaria, Ascobolus carbonarius*).

III. Species fruiting in abundance on two- and three-year-old burns (*Peziza praetervisa, Trichophaea hemisphaerioides, Geopyxis rehmii*).

IV. Species connected with bryophytes, appearing in the third year (*Octospora leucoloma*).

From the observations of different authors it follows that the succession of fungi on fireplaces both in the lowland and in the mountains in different countries of Europe is similar.

6.3.3. *Carbophilous mycocoenoses*

On fire places fungi form fungal communities on the remains of more or less burnt and charred wood. These communities of fungi do not belong to the phytocoenoses of vascular plants within which they occur, although they can have some influence on the development of the fungi, e.g. by the creation of a moist microclimate in the forest. However, the basic factor influencing the development of the mycosociety is the presence of charred wood. Most frequently the initiator of the fireplace community is men and his activity.

Darimont (1973) included the mycocoenoses of fireplaces occurring in different forest associations (Fagetum boreoatlanticum, Quercetum sessiliflorae, Querco-Carpinetum, Aceri-Fraxinetum) to one class: Anthracobietea. The subdivision of this class, modified by Kreisel (in Michael *et al.* 1981) is (Table II):

Class Anthracobietea Darimont 1973
Forest carbophilous fungal communities on fireplaces.
Ch.: *Tephrocybe ambusta, Pholiota* (*Flammula*) *carbonaria, Hebeloma anthracophilum, Coprinus angulatus* (= *C. boudieri*), *C. friesii, Mycena galopus* var. *nigra*.

Order Anthracobio-Flammuletea carbonariae Darimont 1973
Ch.: species of the class.

Alliance Anthracobion melalomae Darimont 1973

Table II. Syntaxonomic differentiation of carbophilous macrofungi

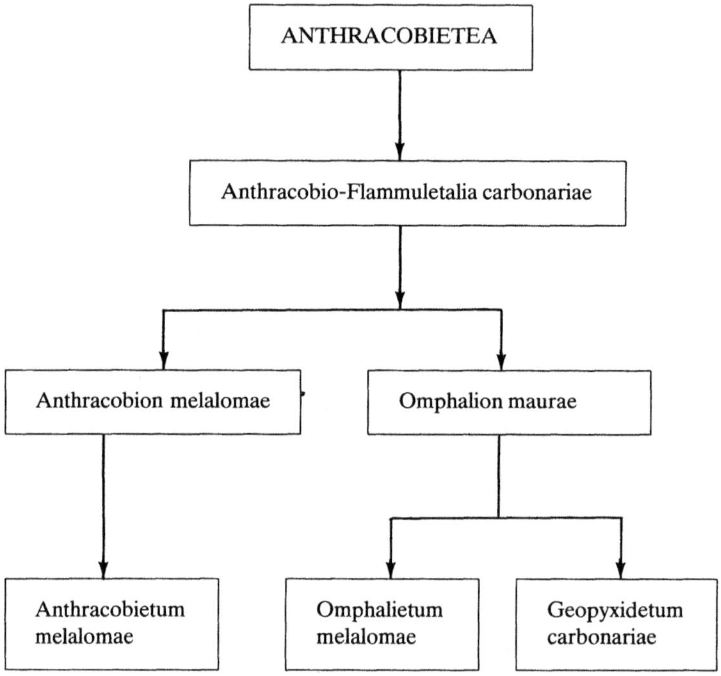

Ch.: Anthracobia melaloma.

Ass. Anthracobietum melalomae Darimont 1973

Anthropogenic community on charcoal in calciphilous forests of Belgium (Aceri-Fraxinetum, Querco-Carpinetum).

Ch.: *Anthracobia melaloma.*

Alliance Omphalion maurae Darimont 1973

Ch.: *Fayodia* (*Omphalia*) *maura, Inocybe lacera* f. *anthracophila, Geopetalum carbonarium, Aleuria aurantia.*

Ass. Omphalietum maurae Darimont 1973.

Anthropogenic community on charcoal in acidophilous forests of Belgium (Quercetum sessiliflorae). It comprises a mountain variant (in Fagetum boreoatlanticum) with *Rhizina inflata* as distinguishing species.

Ch.: species of the alliance.

Ass. Geopyxidetum carbonariae Ebert 1958

On the remains of charred wood in coniferous forests of Germany, Poland. Also the community of *Geopyxis carbonaria* and *Aleuria carbonaria* can be grouped to it. It was observed by Moser (1949) in a recently burnt spruce forest in Austria. The table of macrofungi listed by Pirk (1950) on fireplaces in moss-communities of West Germany belongs very likely to this association.

Ch.: *Geopyxis carbonaria, Peziza violacea, Pholiota carbonaria.*

6.4. Coprophilous macrofungi

6.4.1. *Germination conditions and relations between coprophilous fungi and different kinds of dung*

Coprophilous fungi are a group of saprophytic fungi living on animal excrements. From the taxonomic point of view they mainly belong to several families of Pezizales and Agaricales.

The literature referring to the taxonomic problems of this fungal group is rather rich (Boudier 1869, Massee and Salmon 1901, 1902, Lange 1952, Brummelen 1967, Kimbrough 1965, 1966, 1969, Eckblad 1968, Larsen 1970, and others). Many publications have a floristic character (Chełchowski 1892, Moravec 1968, Binyamini 1972, Kalamees 1980, and others).

Studies on the biology of coprophilous fungi were carried out, among others, by Boudier (1869), Janczewski (1871), Massee and Salmon (1901), Fraser (1907), Dodge (1912), Larsen (1970). These studies refer primarily to the methods of spreading and spore germination of the investigated fungi.

Depending on the conditions of spore germination the coprophilous fungi have been divided in the following way (Larsen 1971):

I. Endocoprophilous fungi

   Spores can germinate after passing the alimentary tract of herbivorous animals (the majority of coprophilous fungi) — Figure 1.

   (a) Obligate endocoprophilous fungi

      Spores germinate only after passing the alimentary tract of animals (e.g. species of *Ascobolus* and *Saccobolus*)

   (b) Facultative endocoprophilous fungi

      Spores can germinate also without passing the alimentary tract of animals (e.g. *Arachniotus, Coprinus*)

II. Ectocoprophilous fungi

   Spores cannot germinate after passing the alimentary tracts of animals (probably *Ascozonus*)

Arnolds (1981) distinguishes the proper (obligate) coprophilous fungi growing exclusively on dung e.g. *Saccobolus glaber, Anellaria semiovata, Stropharia semiglobata, Psilocybe coprophila*, and subcoprophilous fungi

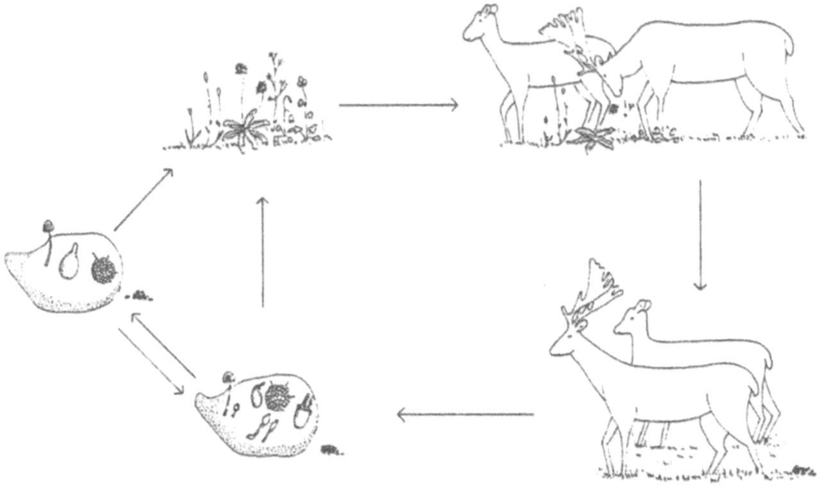

Fig. 1. Spore dispersal in coprophilous fungi (acc. to Larsen 1971).

(facultative coprophytes) growing also on fertile habitats, e.g. *Bolbitius vitellinus, Conocybe siliginea, Panaeolus fimicola, P. acuminatus, Panaeolina foenisecii.* On grasslands the intensity of pasturing by cattle, and particularly by cows, influences positively abundant occurrences of coprophilous fungi.

Coprophilous fungi are connected in a variable degree with the dung of different animals. This problem was dealt with in an extensive way by Richardson (1972). He investigated the occurrence of coprophilous Ascomycetes on 137 dung samples of horses, sheep, cows, roe-deers, stags, rabbits, and hares collected in England, Ireland, Jugoslavia, and New Zealand. The occurrence of the less frequent fungi on the dung of different animals is shown in Table III.

Some coprophilous fungi occur on different types of dung, e.g. *Thelebolus nanus, Podospora vesicola, Saccobolus versicolor, Coprinus miser, C. heptemerus*; other coprophytes show a close preference for one type of dung, e.g. *Podospora setosa, Psilocybe coprophila* on rabbit dung; *Coprinus ephemeroides, Stropharia semiglobata, Ascobolus immersus, Cheilymenia pulcherrima* on sheep dung; *Coprobia granulata* on cow dung. Arnolds (1981) distinguishes also fungi species dominating on rabbit, horse and sheep excrements, e.g. *Coprinus cordisporus* and *Lasiobolus ciliatus*, and species dominating on cow and sheep excrements, e.g. *Coprinus velox, Coprobia granulata*, and *Saccobolus glaber.*

The dung of exotic animals was investigated already earlier (Massess and

Table III. Occurrence of less frequent coprophilous fungi on different dung types (acc. to Richardson 1972)

	Horse	Sheep	Cow	Roe deer	Rabbit	Hare	Total no.
No. of samples	20	36	20	19	32	10	137
Saccobolus versicolor (Karst.) Karst.	4	15	2	2	10	0	33
Ascophanus carneus (Pers.) Boud.	4	14	6	2	5	0	31
Sphaeronaemella fimicola Marchal	1	8	0	8	10	1	28
Ascobolus stictoideus Speg.	0	9	2	6	4	1	22
Ascophanus microsporus (Berk. and Br.) Hansen	0	9	5	7	1	0	22
Ascobolus furfuraceus Pers. ex Fr.	1	2	9	7	3	0	22
Rhyparobius polysporus (Karst.) Speg.	3	7	1	2	6	2	21
Coniochaeta scatigena (Berk. and Br.) Cain	1	7	6	0	2	4	20
Phomatospora coprophila Richardson	0	14	2	2	0	0	18
Theleborus stercoreus Tode	0	2	0	4	9	3	18
Sporormia ambigua Niessl	5	3	3	1	4	2	18
Podospora appendiculata (Auersw.) Niessl	1	3	1	0	4	7	16
Coprobia granulata (Bull. ex Fr.) Boud.	0	4	8	4	0	0	16
Trichodelitschia bisporula (Crouan) Hansen	1	10	1	0	0	3	15
Cheilymenia coprinaria (Cooke) Boudier	1	9	2	1	0	0	13
Coniochaeta discorspora (Auersw.) Cain	1	2	2	0	4	4	13
Podospora setosa (Winter) Niessl	1	1	0	2	8	1	13
Ascozonus woolhopensis (Renny) Schroet.	0	0	0	2	9	0	11
Sporormia minima Auersw.	0	2	5	0	2	1	10
Rhyparobius sexdecimsporus (Crouan) Sacc.	3	6	0	1	0	0	10
Ascobolus crenulatus Karst.	0	2	0	1	5	1	9
Sporormia gigantea Hansen	1	4	2	0	0	1	8
Coniochaeta hansenii (Oud.) Cain	0	0	1	0	2	5	8
Sordaria fimicola (Rob.) Ces. and de Not.	2	2	0	1	1	0	6
Sporormia bipartis Cain	0	0	0	0	6	0	6
Cheilymenia stercorea (Pers.) Boud.	2	0	4	0	0	0	6
Saccobolus glaber (Pers.) Lamb.	2	1	0	0	0	0	3

Salmon 1901) but it was sampled in the local zoological gardens, therefore both the food and the habitat were artificial.

Stoll (1934) pointed to the attachment of some coprophilous fungi to the structure (consistency) of the dung and he distinguished two types of dung:

I. Dry, compact excrements (pellets) with a high content of poorly decomposed material from rabbits and horses,

II. pulpy, soft, well decomposed cow dung.

Sheep dung can belong to both mentioned types depending on the food consumed by the sheep, e.g. in case of poor heath vegetation the sheep dung is dry in the form of pellets, on the other hand when sheep graze on rich vegetation, their dung has a consistency resembling cow dung. Hence Stoll (l.c.) could cultivate on sheep dung all coprophilous fungi found on different types of dung. Arnolds (1981) regards horse dung as equally rich in different coprophilous fungal species, while Kohlman-Adamska (1965) found that cow dung is characterized by almost twice as rich coprophilous flora than the dung from goat or horse.

Summing up, one can find that the occurrence of some species of coprophilous fungi on animal dung depends on three factors (Richardson 1972):

I. Physical character of the dung, its consistency, moisture, holding capacity.

II. Chemical characters of the dung.

III. Biological character of the dung, the influence of other organisms developing on the investigated dung, and the mutual relations (association and antagonism) between the different groups of coprophilous fungi.

6.4.2. *Succession of macrofungi on dung*

The successions of coprophilous fungi are commonly known, although there are comparatively few elaborations on this subject. They can be observed in laboratory conditions on fresh animal dung placed in a moist chamber (Stoll 1934, Harper and Webster 1964, Kohlman-Adamska 1965, Larsen 1970, Mitchel 1970, Singh 1984).

The authors agree that in the first stage of succession, about 24 hours after the establishment of the culture, there appear coprophilous Zygomycetes (mainly species of *Pilobolus, Mucor* and *Kickxella*). Kohlman-Adamska (1965) calls this stage the aspect of Phycomycetes. In the second stage of succession, after several days, there develop Ascomycetes, first Discomycetes, mainly Ascobolaceae. Species of *Ascophanus, Ascobolus* and *Saccobolus* are characterized by a quick development (5—8 days). This stage is called by Kohlman-Adamska the aspect of Ascobolaceae. In their development Ascobolaceae outpace Pyrenomycetes, representatives of Lasiosphaeriales (*Sordaria, Podospora, Pleurage, Bombardia*), which develop in 9—14—30 days. The period of their maturation can be called the aspect of Lasiosphaeriaceae. The succession of coprophilous fungi is terminated by the occurrence of Basidiomycetes, mainly species of *Coprinus* and *Stropharia*. This stage is called by Kohlman-Adamska the aspect of *Coprinus*.

It must be stressed, that the above described succession of coprophilous fungi is the succession on fruit-bodies only. The succession on dung observed in field conditions has a much slower course than in laboratory conditions, probably due to the fluctuations of temperature, moisture, and water content in the substrate in natural conditions.

6.4.3. *Coprophilous fungal communities*

Pirk and Tüxen (1949) were the first who described communities of coprophilous fungi on horse and cow excrements in natural conditions on grassland in Niedersachsen (Germany). This community type has been defined as an association Coprinetum ephemeroidis Pirk and Tüxen 1949 on the basis of an analysis of 16 mycosociological records made on plots of 10—100 m^2 situated in places abunding in dung. The majority of the plots was investigated only once. The fungal species found by the authors were tabulated and they distinguished 21 characteristic species (*Coprinus ephemeroides, Bolbitius vitellinus, B. titubans, Panaeolus leucophanes, P. subbalteatus, P. papilionaceus, Psilocybe coprophila*, and others).

Pirk observed the succession of Coprinetum ephemeroidis on cow dung in a period of four months. On fresh dung after 5 days the fructification of *Coprinus ephemeroides* appeared first, after 14 days *Coprinus narcoticus* and *Bolbitius titubans* showed up, after 23 days there developed *Psilocybe merdaria* and other species, after 29 days *Stropharia stercoraria*, and after 47 days the cow dung was decomposed for 9/10 and overgrown by grasses, and non-coprophilous meadow fungi started showing up.

Svrček (1960) described associations of coprophilous fungi on 4 types of dung: of horse, cow, roe-deer, and dog, and he presented lists of fungi species occurring on the investigated dungs.

Wojewoda (1975) distinguished a community type of coprophilous fungi on animal excrements in Ojcowski National Park (Poland). He gave it the rank of an independent association and called it Strophorietum semiglobatae. As the main characteristic species Wojewoda recognized *Stropharia semiglobata*, and furthermore: *Anellaria semiovata, Coprinus cinereus, C. hiascens, C. patouillardii, C. radiatus, Psilocybe coprophila*, and *Panaeolus sphinc-*

trinus. The described community type, according to the author, cannot be regarded as synusia within the associations of vascular plants, although the 'macro-community' exerts some influence on fungi growing on accidental excrements in this community. They will develop better on a forest road in the shadow of trees, and the development will be worse on open meadows.

Barkman (1976b) distinguished 4 synusiae of coprophilous fungi in the brushwood of *Juniperus* on rabbit dung and on the dung of man, dog, cow, and horse. The author did not give any names to the particular synusiae.

A syntaxonomic classification of coprophilous fungi proposed by Fellner (1988) is presented in Table IV.

6.4.4. *Periodicity of the fructification of coprophilous fungi*

The problem of the seasonal fructification of macrofungi is discussed in many mycosociological works, however in relation to coprophilous fungi this question is usually omitted.

Svrček (1959) found the months of winter and spring as the optimal fruiting period for the majority of species.

According to the observation by Kohlman-Adamska (1965) the autumn and then the spring were found to be the most favourable seasons of the year for the development of coprophilous Ascomycetes cultured from the surroundings of Warsaw. Differences in the species composition were found as well. *Lasiobolus ciliatus* fructified only in spring, *Saccobolus kerverni* and *Pleurage zygospora* only in summer, and *Ascophanus holmskijoldii, Saccobolus depauperatus*, and *Pleurage fimiseda* fruited only in autumn. Some species occurred during the whole vegetative period.

6.5. Macrofungi on arthropods

Several species of insects are affected by parasitizing fungi which show a far advanced specialization in the choice of hosts. Such entomogenic parasites comprise primarily Entomophthoraceae and Laboulbeniales, as well as some Deuteromycetes, e.g. *Paecilomyces farinosus* (a known parasite of many insects (Fassatiova 1955, 1956).

Parasitizing macrofungi occurring on pupae of insects, and sometimes on spiders and flies never create any major groups (mycosocieties), but they are spread in different plant communities. They are lacking only in very dry and light grassland communities like Spergulo-Corynephorion and strongly fertilized Agropyro-Rumicion (Arnolds 1981).

Table IV. Syntaxonomic differentiation of coprophilous macrofungi

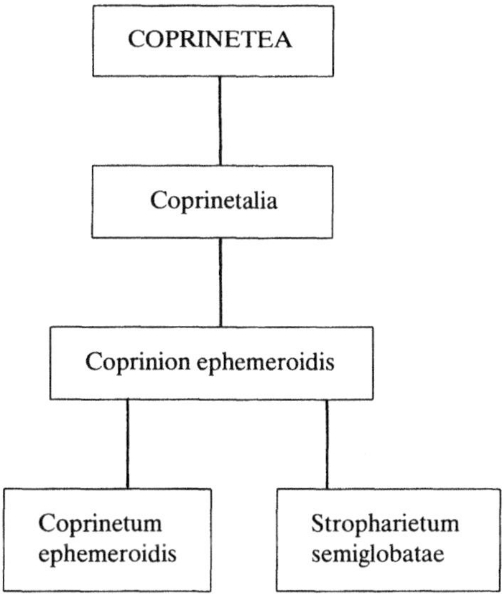

The macrofungus most frequently observed on insects is *Cordyceps militaris*. It develops in caterpillars buried in the litter or on pupae of butterflies, cockchafers, and hymenoptera causing the death of the insect. A list of the hitherto known hosts has been presented by Kobayasi (1941). It is a cosmopolitan fungus known from different climatic zones of North and South America, Asia and Europe. It is found both in moist forests (Carici-Alnetum, Circaeo-Alnetum), eutrophic deciduous forests and mixed forests (Dentario glandulosae-Fagetum, Querco-Carpinetum) as well as in coniferous forests (Pino-Quercetum, Piceetum tatricum, Leucobryo-Pinetum) in the lowland and in the mountains (Komorowska 1981). Barkman (1976b) regards *Cordyceps militaris* as a distinguishing species for Sqarroso-Juniperetum and Dicrano-Juniperetum, preferring eutrophic habitats. *Cordyceps norvegica* mentioned by Eckblad (1967) is presumably only a synonym of *Cordyceps militaris*. Among other species of *Cordyceps* developing on insects Eckblad mentions *Cordyceps entomorrhiza* found on larvae and imagines of *Carabus*. A related species is *Cordyceps gracilis* found on larvae of Lepidoptera (Dennis 1978). *Cordyceps bifusispora* was described on a pupa of Noctuidae in Sweden (Eriksson 1982). Furthermore *Cordyceps pistillariaeformis* was found among others on plant louses of *Lecanium* in North America and England (Massee 1895). *Cordyceps sphingum* is known almost all over the world on butterflies of *Sphinx* genus (Zabłocka 1929).

Munk (1957) mentions species found in Denmark like *Cordyceps cinereus* on dead larvae and imagines of Carabidae, and *Cordyceps myrmecophilus* on dead imagines of *Formica rufa*.

Interesting are exotic termitophilous Agaricales which Heim (1958) separated into a new genus of *Termitomyces*. They occur in the termitaries of *Termes vulgaris, T. badius* and others. Hereto belong the following species of fungi described from Congo: *Termitomyces microcarpus, T. striatus, T. clypeatus, T. robustus, T. globulus, T. schimperi, T. mammiformis*, and *T. letestui*.

Fellner (1988) proposed to separate the insect parasitizing macrofungi into the class Cordyceptea nom. prov.

6.6. Parasitic macrofungi on fruit-bodies of other fungi

The phenomenon of fungi parasitizing on other fungi is rather frequent, although not sufficiently scientifically investigated. The literature is comparatively scant and refers to observations of fungi parasitizing only on fruit-bodies, and not on their mycelium.

One of the first authors who called attention to the phenomenom of parasitization of fungi on other fungi was Schröter (1886).

Mycoparasites belong to different classes, e.g. in the Zygomycetes a rather numerous group of parasitizing species is shown in the Mucorales, in Ascomycetes in the order of Clavicipitales, and in the Basidiomycetes they belong to the Boletales, Agaricales, and Tremellales.

Most of the mycoparasites show an attachment only to one host genus. On hypogeous fruit-bodies of *Elaphomyces* sp. div. both in deciduous and coniferous forests *Cordyceps ophioglossoides* is most frequently found (Munk 1957, Kobayashi and Shimitau 1960, Eckblad 1967, Dörfelt and Conrad 1978, and others). In Europe on *Elaphomyces* parasitize two *Cordyceps* species, macroscopically similar to each other, i.e. *Cordyceps capitata* and *C. canadensis*, and a very rare species: *Cordyceps japonica* (Mains 1957, Ulvinen 1966, Eckblad 1967, Sałata and Ostas 1975).

Xerocomus parasiticus is an obligate parasite of the fruit-bodies of *Scleroderma citrinum*, and rarely also of *Scleroderma verrucosum* (Šmarda 1963, Skirgiełło 1965, Lange 1974). This fungus grows usually in groups of several to about 20 specimens at the base of the fruit-bodies of *Scleroderma*. It occurs on forest meadows, in acidophilous forest communities on sandy soils, in places where there is abundant light and no rich vegetation. It was recorded in Europe (Lange 1974), America (Singer 1965, Snell and Dick

1970) and in Africa (Singer 1965). A related species, *Xerocomus astraeicola*, occurs in Japan (Imazeki and Hongo 1975).

Asterophora lycoperdoides and *A. parasitica* can be met only on decaying fruit-bodies of *Russula*, less frequently on fruit-bodies of *Lactarius vellereus*.

Many species parasitizing on fungi comprise the genus *Tremella*. The parasitic species of *Tremella* can be connected into groups depending on the systematic affiliation of the host (Wojewoda 1975a):

I. Parasites of Pyrenomycetes: *Tremella globospora* on stroma of *Diaporthe* and *Eutypella, Tremella pyrenophila* on *Valsaria insitiva.*

II. Parasites of Dacrymycetaceae.

III. Parasites of Stereaceae and Corticiaceae: *Tremella encephala* parasitizes on *Stereum sanguinolentum* mainly in Pinetum mughi carpaticum, Piceetum tatricum, Vaccinio myrtilli-Pinetum, and Pino-Quercetum. *Tremella steidleri* most likely a parasite of *Stereum gausapatum, Tremella aurantia* and *T. tremelloides* on Stereum on deciduous trees in North America; *Tremella mycophaga* and *T. simplex* are parasites of *Aleurodiscus amorphus* which grows on *Abies* and *Picea*; *Tremella versicolor* parasitizes on the fruit-bodies of different species of Corticiaceae.

IV. Parasites of Polyporaceae: *Tremella polyporina* on *Tyromyces caesius* and *T. lacteus.*

In the conclusion some mention must be made about *Collybia cirrhata, C. cookei* and *C. tuberosa* developing saprophytically on decaying fruit-bodies of *Paxillus, Lactarius* and *Russula*, but also on decaying litter in deciduous forests.

6.7. Muscicolous macrofungi

Several macrofungi are more or less closely associated with mosses. They grow either on living shoots of some moss species, or they create fruit-bodies among mosses which inhabit different substrates, e.g. decaying stumps, the base of living tree trunks, or moist soil.

In mycosociological papers some authors distinguish muscicolous macrofungi as a special ecological group. It may serve for better and more complete characteristics of the habitats in forest associations. And so for example Bujakiewicz (1981, 1982) says that macrofungi developing among mosses and peatmosses are well-distinguishing species in the associations of Sphagnetum magellanici and Bazzanio-Piceetum (e.g. *Galerina paludosa, G. tibiicystis*, and others). Muscicolous fungi as *Galerina hypnorum, G. sahleri*, and *G. mniophila* are also distinguishing for the Abieti-Piceetum montanum and Piceetum excelsae carpaticum in the massif of Babia Góra in the belt of West Carpathians.

Arnolds (1981, 1982) mentions a number of macrofungi associated with mosses in meadow associations of Drenthe, the Netherlands, among others *Galerina vittaeformis, G. hypnorum, G. heterocystis, G. laevis, Geoglossum nigritum, Octospora humosa, Rickenella fibula, R. setipes, Mycenella bryophila, Omphalina pyxidata*. Several species of macrofungi always accompany peatmosses (*Sphagnum*), such as *Hypholoma udum, H. elongatipes, Galerina paludosa, G. sphagnorum, Omphalina sphagnicola, Tephrocybe palustris* (Lange 1948, Kreisel 1957).

References

Arnolds, E. 1981, 1982. Ecology and coenology of macrofungi in grasslands and moist heathlands in Drenthe, the Netherlands. Bibl. mycol. 83, 90.

Arnolds, E. 1988. Status and classification of fungal communities. p. 153—165. *In*: J. J. Barkman and K. V. Sykora (eds.) Dependent plant communities. SPB Academic Publishing, The Hague.

Babos, M. 1981. Mycological examination of sawdust depots in Hungary. Studia Botanica Hungarica 15: 31—44.

Barkman, J. J. 1973. Synusial approaches to classification. Handbook of Vegetation Science 5: 435—491.

Barkman, J. J. 1976a. Algemene inleiding to de oekologie en sociologie van macrofungi. Coolia 19: 57—66.

Barkman, J. J. 1976b. Terrestrische fungi in jeneverbesstruwelen. Coolia 19: 94—110.

Barkman, J. J. 1987. Methods and results of mycocoenological research in the Netherlands. p. 7—38. *In*: G. Pacioni (ed.) Studies on fungal communities. University of L'Aquila, Italy.

Barkman, J. J., A. E. Jansen and B. W. L. de Vries, 1983. De betekenis van dood hout voor de schimmelflora. Ned. Bosbowtijdschrift 55: 57—64.

Binyamini, N. 1972. Some coprophilous fungi of Israel, 1. Ascomycetes. Israel Journal of Botany 21: 112—116.

Boudier, E. 1869. Mémoire sur les Ascobolés. Ann. Sci. Nat. Bot. 10: 191—268.

Brummelen, J. van. 1967. A world-monograph of the genera *Ascobolus* and *Saccobolus* (Ascomycetes, Pezizales). Persoonia, Suppl. 1.

Bujakiewicz, A. 1981, 1982. Grzyby Babiej Góry, II, III. Acta Mycol. 17: 63—125, 18: 3—44.

Butin, H., and Kappich, I. 1980. Untersuchungen zur Neubesiedlung von verbrannten Waldböden durch Pilze und Moose. Forstw. Cbl. 99: 283—296.

Chełchowski, S. 1982. Przyczynek do znajomości krajowych grzybów gnojowych (Fungi fimicoli Polonici). Pam. Fizjogr. 12: 171—179.

Darimont, F. 1973. Recherches mycosociologiques dans les forêts de Haute Belgique. Inst. Roy. Sci. Nat. Belg. Mémoire 170.

Dennis, R. W. G. 1981. British Ascomycetes. J. Cramer, Vaduz.

Dodge, B. O. 1912. Methods of culture and the morphology of the archicarp in certain species of the Ascobolaceae. Bull. Torrey Bot. Club 39: 139—197.

Dörfelt, H. 1977. Beiträge zur Kenntnis der Naturschutzgebiete in den Bezirken Halle und Magdeburg IV. Naturschutz u. naturkundl. Heimatforschung i. d. Bez. Halle u. Magdeburg 14: 49—55.

Dörfelt, H., and Conrad, R. 1978. Die Kernkeulen (*Cordyceps*-Arten) in Ostthüringen und im sächsischen Vogtland. Veröff. Mus. Gera Naturwiss. 6: 41—52.

Ebert, P. 1958. Das Geopyxidetum carbonariae, eine carbophile Pilzassoziation. Zeitschr. f. Pilzkunde 24: 32—44.

Eckblad, F.-E. 1967. The genus *Cordyceps* in Norway. Nytt Magasin for Bot. 14: 68—76.

Eckblad, F.-E. 1968. The genera of the operculate Discomycetes. Nytt Mag. Bot. 15: 1—184.

Einhellinger, A. 1977. Die Pilze in primären und sekundären Pflanzengesellschaften oberbayerischer Moore. II. Ber. Bayer. Bot Ges. 48: 61—146.

Eriksson, O. 1982. *Cordyceps bifusispora* spec. nov. Mycotaxon 15: 185—188.

Fassatiova, O. 1955. O isáriových formách entomofágnich hub. Čes. Mykol. 9: 134—139.

Fassatiova, O. 1956. *Spicaria farinosa* (Dicks) Vuil. na puklici jasanové. Čes. Mykol. 10: 242—245.

Favre, J. 1955. Les champignons supérieurs de la zone alpine du Parc National Suisse. Ergebn. wiss. Unters. Schweiz. Nat. Parks 5 (N. F.) 33: 1—212.

Favre, J. 1960. Catalogue descriptif des champignons supérieurs de la zone subalpine du Parc National Suisse. Ergebn. wiss. Unters. Schweiz. Nat. Parks 6 (N.F.) 42: 321—610.

Fellner, R. 1988. Poznámky k mykocenologické syntaxonomii. 2. Česká Mykol. 42:41—51.

Fraser, H. C. I. 1907. On the sexuality and development of the ascocarp in *Lachnea stercorea* Pers. Ann. Bot. (London) 21: 349—360.

Grabherr, W. 1936. Die dynamik der Brandflächenvegetation auf Kalk- und Dolomitböden des Karwendels. Beih. Bot Centralbl. 55: 1—94.

Harper, J. E., and J. Webster 1964. An experimental analysis of the coprophilous fungus succession. Trans Brit. Mycol. Soc. 47: 511—530.

Heim, R. 1958. Flore iconographique des champignons du Congo, 7: Termitomyces. Bruxelles.

Imazeki, R., and T. Hongo. 1975. Coloured illustrations of fungi of Japan 1. Hoikuska Publish Co., Ltd., Osaka.

Jahn, H. 1962. Pilzbewuchs an Fichtenstümpfen (*Picea*) in westfälischen Gebirgen. Westfäl. Pilzbriefe 3: 110—122.

Jahn, H. 1965. Die *Phellinus robustus* var. *hippophaës-Ph. contiguus*-Assoziation, eine Pilzgesellschaft an Sanddorn. Westf. Pilzbriefe 5: 139—140.

Jahn, H. 1966. Pilzgesellschaften an *Populus tremula*. Zeitschr. f. Pilzkunde 32: 26—42.

Jahn, H. 1968. Das Bisporetum antennatae, eine Pilzgesellschaft auf den Schnittflächen von Buchenholz. Westfäl. Pilzbriefe 7: 41—47.

Jahn, H. 1979. Pilze die an Holz wachsen. Busse, Herford.

Jahn, H. 1990. Pilze an Bäumen. 2. edition revised by H. Reinartz and M. Schlag. Patzer, Berlin and Hannover.

Janczewski, E. 1871. Morphologische Untersuchungen über *Ascobolus furfuraceus*. Bot. Zeitung (Berlin) 29: 257—262.

Käärik, A. A. 1974. Decomposition of wood. p. 129—174. *In*: C. H. Dickinson and G. J. F. Pugh Biology of plant litter decomposition. London and New York.

Kalamees, K. 1980. Trophic groups of Estonian agarics. Ecology and distribution of fungi: 71—98.

Keizer, P.-J. 1985. Oecologie en taxonomie van hout-bewonende Aphyllophorales in moerasbossen in Drente. Landbouwhogeschool, Utrecht.

Kimbrough, J. W. 1965. Studies in the Pseudoascoboleae. Canad. J. Bot. 44: 685—704.

Kimbrough, J. W. 1966. The structure and development of *Trichobolus zukalii*. Mycologia 58: 289—306.

Kimbrough, J. W. 1969. North American species of *Thecotheus* (Pezizeae, Pezizaceae). Mycologia 61: 99—114.

Kobayasi, Y. 1941. The genus *Cordyceps* and its allies. Sci. Rep. Tokyo Bunr. Daig. 5: 53—260.

Kobayasi, Y., and D. Shimitsu 1960. Monographic studies of *Cordyceps* 1. Group parasitic on *Elaphomyces*. Bull. Nat. Sci. Mus. (Tokyo) 5: 69—85.

Kohlman-Adamska, A. 1965. Niektóre grzyby koprofilne z okolic Warszawy. Acta mycol. 1: 77—103.

Komorowska, H. 1981. *Cordyceps militaris* (Vaill. ex L.) Link (Ascomycotina) w Polsce. Fragm. Flor. et Geobot. 27: 657—666.

Kotlaba, F. 1984. Zeměpisné rozšiřeni a ekologie chorošů (Polyporales s. 1.) v Československu, Academia, Praha.

Kotlaba, F. 1985a. Ekologie a rozšiřeni pevniku význačného — *Stereum insignitum* se zvláštnim zřetelem k Československu. Česka Mykologie 39: 1—14.

Kotlaba, F. 1985b. Pozaruhodný pevnik *Stereum subpileatum* (Aphyllophorales), jeho ekologie a rozšiřeni se zvláštnim zřetelem k Československu. Česka Mykologie 39: 193—204.

Kreisel, H. 1957. Die Pilzflora des Darß und ihre Stellung in der Gesamtvegetation. Fedd. Rep. Beih. 137: 110—183.

Kreisel, H. 1961. Die Entwicklung der Mykozönose an *Fagus*-Stubben auf norddeutschen Kahlschlägen. Fedd. Rep. 139: 227—232.

Kreisel, H. 1971. Charakteristik der Pilzflora Kubas. Biol. Rundschau 9: 65—73.

Kreisel, H., and K. H. Müller. 1987. Das Pleurotetum cornucopiae, eine Pilzgesellschaft an toten Ulmenstämmen im Gefolge des Ulmensterbens. Arch. Nat.schutz Landsch. forsch. 27: 17—25.

Lange, L. 1974. The distribution of macromycetes in Europe. Dansk Bot. Ark. 30: 1—105.

Lange, M. 1948. The agarics of Maglemose. Dansk Bot. Ark. 13: 1—141.

Lange, M. 1952. Species concept in the genus *Coprinus*. Dansk Bot. Ark. 14: 1—164.

Larsen, K. 1970. The genus *Saccobolus* in Denmark. Bot. Tidsskr. 65: 371—389.

Larsen, K. 1971. Danish endocoprophilous fungi, and their sequence of occurrence. Bot. Tidsskr. 66: 1—32.

Lisiewska, M. 1974. Macromycetes of beech forests within the eastern part of the *Fagus* area in Europe. Acta Mycol. 10: 3—72.

Mains, E. B. 1957. Species of *Cordyceps* parasitic on *Elaphomyces*. Bull. Tor. Club 84: 243—251.

Martin, S. W. 1952. Revision of the North Central Tremellales. State Univ. Iowa Stud. Nat. Hist. 19: 1—122 (reprint 1969).

Massee, G. 1895. A revision of the genus *Cordyceps*. Ann. of Bot. 9: 1—45.

Massee, G., and E. S. Salmon. 1901. Researches on coprophilous fungi. Ann. Bot. 15: 313—338.

Massee, G., and E. S. Salmon. 1902. Researches on coprophilous fungi II. Ann. Bot. 16: 57—93.

Michael, E., B. Hennig, and H. Kreisel. 1981. Handbuch für Pilzfreunde, Vol. IV. G. Fischer, Jena.

Mitchel, D. T. 1970. Fungus succession on dung of south African Ostrich and Angora Goat. J. S. Afr. Bot. 36: 191—198.

Moravec, Z. 1968. Remarks on some coprophilous fungi in Norway. Česká Mykologie 22: 301—309.

Moser, M. 1949. Untersuchungen über den Einfluß von Waldbränden auf die Pilzvegetation I. Sydowia 3: 336—383.

Munk, A. 1957. Danish Pyrenomycetes. Dansk Bot. Ark. 17: 1—491.

Paulsen, M. D., and H. Dissing, H. 1979. The genus *Ascobolus* in Denmark. Bot. Tidsskr. 74: 67—78.

Petersen, R. M. 1970. Danish fireplace fungi. An ecological investigation on fungi on burns. Dansk Bot. Ark. 27: 1—97.

Pilát, A. 1957. Přehled evropských Auriculariales a Tremellales se zvláštnim zretelem k československým druhům. Dbotn. Nár. Musea v Prace 13B: 115—210.

Pilát, A. 1959. Houby Československa, ve svém životnim prostřeni. Academia, Praha.

Pirk, W. 1950. Pilze in Moosgesellschaften auf Brandflächen. Mitt. Flor.-Soz. Arbeitsgem. N.F. 2: 3—5.
Pirk, W. 1952. Die Pilzgesellschaften der Baumweiden im mittleren Wesertal. Mitt. Flor.-Soz. Arbeitsgem. N.F. 3: 93—96.
Pirk, W., and R. Tüxen. 1949. Das Coprinetum ephemeroidis, eine Pilzgesellschaft auf frischem Mist der Weiden im mittleren Wesertal. Mitt. Flor.-Soz. Arbeitsgem. N.F. 1: 71—77.
Pirk, W., and R. Tüxen. 1957. Das Trametetum gibbosae, eine Pilzgesellschaft modernder Buchenstümpfe. Mitt. Flor.-Soz. Arbeitsgem. N.F. 6/7: 120—126.
Rayner, A. D. M., and L. Boddy. 1988. Fungal decomposition of wood. Chichester.
Ricek, E. W. 1967, 1968. Untersuchungen über die Vegetation auf Baumstümpfen 1, 2. Jahrbuch d. Oberöster. Musealver. 112: 185—252, 113: 229—256.
Richardson, M. J. 1972. Coprophilous Ascomycetes on different dung types. Trans. Br. Myc. Soc. 58: 37—48.
Runge, A. 1975. Pilzsukzession auf Laubholzstümpfen. Zeitschr. f. Pilzkunde 41: 31—38.
Runge, A. 1978. Pilzsukzession auf Kiefernstümpfen. Z. Mykol. 44: 295—301.
Runge, A. 1980. Pilz-Assoziationen auf Holz in Mitteleuropa. Z. Mykol. 46: 95—102.
Runge, A., and F. Runge, F. 1979. Das Trametetum gibbosae in Westfalen. Decheniana 132: 1—2.
Rypácek, V. 1966. Biologie holzzerstörender Pilze. G. Fischer, Jena.
Sałata, B., and T. Ostas. 1975. Nowe stanowiska interesujących grzybów wyższych (macromycetes) w południowo — wschodniej Polsce. Fragm. Flor. et Geobot. 21: 521—526.
Schmitt, J. A. 1987. Zur Ökologie holzbesiedelnder Pilzarten. Aus Natur und Landschaft im Saarland. Sonderband 3: 101—119.
Schröter, J. 1886. Über die auf Hutpilzen vorkommenden Mucorineen. Jahresber. Schles Ges. vat. Kultur 64.
Singer, R. 1965. Die Röhrlinge I. Die Pilze Mitteleuropas 5., J. Klinkhardt, Bad Heilbrunn.
Singh, C. S. 1984. Successional studies of fungi on mammalian dung. Acta Mycol. 20: 105—120.
Skirgiełło, A. 1965. Materiały do poznania rozmieszczenia geograficznego grzybów wyższych w Europie. I. Acta Mycol. 1: 23—26.
Snell, W. H., and E. A. Dick. 1970. The Boleti of Northeastern North America. J. Cramer, Lehre.
Stoll, K. 1934. Untersuchungen über die koprophilen Pilze unserer Haustiere. Zbl. Bakt. II. Abt. 90: 97—127.
Strid, A. 1975. Wood-inhabiting fungi of alder forests in North-Central-Scandinavia. Wahlenbergia 1.
Svrček, M. 1959. Nekolik zajimavych druhu koprofilnich hub pozorovanych v roce 1958. Česká Mykologie 13: 92—100.
Svrček, M. 1960. Eine mykofloristische Skizze der Umgebung von Karlštejn (Karlstein) in Mittelböhmen. Česká Mykologie 14: 67—83.
Šmarda, F. 1963. Přispevěk k mapováni makromycetů v Evropě na přikladu suchohřibu přiživného — *Xerocomus parasiticus* (Bull. ex Fr.) Quél. v Československu. Česká Mykologie 17: 127—133.
Tortić, M. 1962. Primjer sukcessije kod viših gljiva. Acta Botanica Croatica 20/21: 199—202.
Tortić, M. 1981. Istraživanja poliporoidnih i korticioidnih gljiva (Basidiomycetes) u Jugoslaviji. Acta Biol. Iugoslavica, Biosistematika 7: 1—9.
Tortić, M. 1985. Non-poroid lignicolous Aphyllophorales (Fungi, Basidiomycetes) in the Plitvička Jezera National Park (Yugoslavia). Acta Biol. Iugoslavica, Biosistematika 11: 1—15.
Tortić, M., and M. Karadelev. 1986. Lignicolous macromycetes in the submediterranean part of Macedonia (Yugoslavia). Acta Bot. Croat. 45: 109—117.

Turnau, K. 1984. Post-fire cup-fungi of Turbacz and Stare Wierchy Mountains in the Gorce Range (Polish Western Carpathians). Zeszyty Nauk. Uniw. Jagiel. Prace Bot. 12: 145—170.

Ulvinen, T. 1966. *Cordyceps canadensis* Ell. and Ev. in Finnland. Lounais-Hämeen. Luonto 23: 60—63.

Velenovský, J. 1925. České druhy rodu *Leptoglossum* Karst. Mykologia 2: 44—47.

Vries, B. W. L. de, and T. W. Kuyper. 1988. Effect of vegetation type on decomposition rates of wood in Drenthe, The Netherlands. Acta Bot. Neerl. 37: 307—312.

Wałek-Czernecka, A. 1976. Fungi destroying railroad ties in Poland. Translation from Acta Soc. Bot. Pol. Foreign Scientific Publications Department of the National Center for Scientific, Technical and Economic Information, Warsaw.

Winski, A. 1987. Pilzsoziologische Untersuchungen in verschiedenen Waldgesellschaften des südlichen Oberrheingebiets. Diss. Freiburg i. Br.

Wojewoda, W. 1975a. Gatunki rodzaju *Tremella* pasożytujące na grzybach. Wiad. Bot. 19: 119—123.

Wojewoda, W. 1975b. Macromycetes Ojcowskiego Parku Narodowego. II. Acta Mycol. 11: 163—209.

Wojewoda W. 1981. Mała flora grzybów, 2: Basidiomycetes: Tremellales, Auriculariales, Septobasidiales. Państwowe Wydawnictwo Naukowe (State Scientific Publishing House), Warszawa-Kraków.

Zabłocka, W. 1929. O kilku stanowiskach maczużnika (*Cordyceps*). Acta Soc. Bot. Pol. 6: 880—191.

7. The analysis of communities of saprophytic microfungi with special reference to soil fungi

WALTER GAMS

Abstract

The distinction between microfungi and macrofungi reflects ecological differences. Microfungi are usually cosmopolitan and have an enormous potential for dispersal. An efficient enzyme equipment and competitive saprophytic ability characterize the saprophytic mode of life. The present review is mainly concerned with saprophytic soil fungi.

The techniques used for soil-fungal analyses are briefly reviewed. The choice depends on the purposes of the study, such as the microbiological characterization of vegetation and soil types, studies of decomposition processes, etc. This paper is mainly confined to synecological studies of the fungal community; methods of soil sampling, direct observation and isolation techniques are assessed. Parameters for quantification and statistical analysis are subsequently reviewed.

The soil mycoflora comprises species of all major categories of fungi. Besides a nutritional grouping of the individual components, the potential of obligocarbotrophic growth deserves much attention. Specialist groups of thermophilic, heat-resistant, osmophilic, xerophilic and nematophagous fungi are considered. The possibility of chemical elimination of certain functional groups opens an important tool in ecological studies. In a discussion of growth patterns and strategies, r-strategists with short-lived mycelia and abundant sporulation and K-strategists with a slower growing, more persistent mycelium are confronted. Grimes' concept that stress, disturbance and competition also determine the strategies has some impact on soil fungal ecology.

Fungal successions are partly determined by individual capacities to decompose resistant plant cell wall substances, but also by different speed of fructification. A view that K-strategists appear mainly after r-strategists must be considered with caution.

W. Winterhoff (ed.), Fungi in Vegetation Science, 183–223.

Besides soil fungi, the components of the leaf-surface microflora and the microfloras developing in various kinds of litter are mentioned. Factors that affect fungal colonization patterns include soil depth, microhabitats determined by roots and the rhizosphere, moisture and soil atmosphere, temperature, crop plants, seasonal effects, and anthropogenic disturbance and stress. The latter comprise i.a. acidification and ploughing of agricultural soils. Effects of the soil fauna on fungal colonization are increasingly appreciated. Antagonistic effects of other micro-organisms are recognized in many autecological studies.

The question whether there are associations of microfungi in soils comparable to those of plants is critically considered. The quantitatively dominant species have high communality values in the comparison of different biotopes, the infrequent ones show a large amount of unique variation. Few species are known to have an indicator role for particular ecological factors. Only in rare cases mycologists advance beyond the compiling of species lists in the study of successions.

7.1. Introduction

In most soils, fungi are the major component of the soil microflora. This has been assessed both by direct observation of the biomass (e.g. Bååth and Söderström 1980a, Schnürer *et al.* 1985) and by physiological measurements (Anderson and Domsch 1975, 1978a, b, Schnürer *et al.* 1985). Soil fungi comprise saprophytes, mycorrhizal symbionts and parasites. Their roles have been summed up by Christensen (1989). Other important ecological niches for saprophytic fungi include leaves of green plants (the phyllosphere, including the endophytes growing in symptomless leaves), leaves in all phases of decomposition, other plant residues, dung, other fungi, etc.. The litter-decomposing and soil-inhabiting fungi are the topic of the present review. Wood-decaying fungi are excluded, for an excellent review see Rayner and Boddy (1988); other macromycetes are reviewed by Arnolds in Chapter 2. It is impossible to cover the immense topic in a few pages and I shall illustrate certain statements with results from my own field, although, of course, many other sources could be cited equally well. The major reviews (Dickinson and Pugh 1974, Wicklow and Carroll 1981) give much additional information, while the ecology of saprophytic fungi in the broadest sense has been reviewed by Cooke and Rayner (1984).

The distinction between microfungi (fruiting-bodies generally less than 1 mm in diameter) and macrofungi is convenient for ecological considerations. The former are mainly studied in laboratories by specialists while the latter can also be explored by amateurs. This distinction is also reflected in the distribution patterns, which are more restricted in macrofungi, than in the often cosmopolitan microfungi (Pirozynski 1968; with theoretical expansion in Wicklow 1981b). The former (see Chapter 2) usually show rather strong geographical and ecological limitations, even though they may have a saprophytic mode of life (Cooke and Rayner 1984), and an enormous potential for dispersal (Ingold 1971). The microfungi generally behave according to Bass-Becking's (1934) often cited statement "**Everything is everywhere and the environment selects**."

Saprophytic microfungi contrast with plant-parasitic and symbiotic fungi in being rather more independent of host plants. For their saprophytic mode of life they have developed an efficient enzyme equipment and competitive saprophytic ability (Garrett 1956, Cooke and Whipps 1980).

The delimitation between saprophytic and parasitic fungi is by no means sharp. There is a continuum from saprophytic species to minor pathogens (Salt 1979) in the soil microflora. The majority of commonly isolated saprophytic soil fungi affect plant growth adversely rather than positively when tested singly (Domsch and Gams 1968a). Apparently healthy symptomless plants are commonly colonized internally by a range of endophytic fungi (Petrini 1986, Carroll 1986). Senescent stems and leaves of plants are often invaded by weak parasites from the soil (e.g. *Phoma exigua* is found on many plant species, Boerema and Höweler 1967).

7.2. Methodology

More than in any other synecological discipline, the results of soil-fungal analyses depend on the methods used. A major part of this paper therefore provides a selective survey of methods. The aims of microfungal population analyses are usually different from those governing those for macrofungi. The microbiological characterization of vegetation and soil types is less often the aim than studies of decomposition processes (Swift *et al.* 1979) and soil fertility in relation to micro-organisms (Garrett 1963), an assessment of effects of agricultural practice, pollution, etc. on the soil microflora, or the quantification of certain pathogens or antagonists. For each of these aspects different techniques have to be applied. This review is confined to synecological studies of the fungal community.

Direct observation techniques are available as opposed to indirect techniques of isolation. Garrett (1956) epitomized the difference by stating that "with the plate count method one identifies what one cannot see (i.e. *in situ*), whereas with the direct method one sees what one cannot identify". Moreover, direct methods are most suitable to look for sporulation but "indirect techniques can yield more accurate estimates of the true extent of a fungus in a resource" (Kirby *et al.* 1990).

Techniques for studying soil fungi have been reviewed, amongst others, by Warcup (1967), Parkinson (1970), Parkinson *et al.* (1971), and Johnson and Curl (1972). Recent reviews and evaluation of the techniques of assessing soil fungal activities are given by Kjøller and Struwe (1982) and Frankland (1990). Several authors (particularly those cited above) have proposed that techniques should be standardized in order that results obtained from different laboratories may be comparable. In spite of these efforts, differences in handling the fungi and the varying taxonomic views held by authors still make the comparison of investigations of such investigations very difficult.

It is generally agreed that *fungal hyphae* represent the *active phase*, while the *resting stages* (spores, conidia, chlamydospores, sclerotia) are practically inactive. Therefore it is important to distinguish between fungi present as hyphae and as resting stages (Warcup 1955b); this distinction is the major problem encountered with all techniques.

To establish themselves and survive in a soil, fungi must have resting stages and be affected by the generally present *soil mycostasis* (= fungistasis), i.e. spore germination is inhibited without additional supply of nutrients (voluminous literature since Dobbs and Hinson 1953; comprehensive review by Lockwood 1977). Species whose propagules germinate indiscriminately in soil will die out very soon and will not become soil-borne.

7.2.1. *Soil sampling*

Soil samples for the isolation of fungi are taken at random. Several subsamples are usually bulked to form a composite sample (Johnson and Curl 1972). The separate analysis of individual subsamples would be preferable for subsequent statistical analysis, but is usually prevented by the limitations of available labour. When the influence of particular treatments on the mycoflora is to be studied, block designs are the method of choice, with several blocks for each treatment and analysis of composite samples from each block.

The numbers of subsamples must be carefully chosen when a qualitative and quantitative analysis of the species composition is attempted. A considerable number of replicate samples in space and time is more valuable than a single large bulk sample. A considerable number of replications and an individual analysis of sufficient size (see below) both contribute to revealing a fair proportion of the fungal spectrum present.

Homogeneity of a biotope (natural or artifical uniformity of vegetation) is desirable for soil-fungal analyses. Examples of homogeneous biotopes include *Calluna* heaths (Sappa 1955), pine forests without undergrowth (Burges 1963), salt marshes, and agricultural fields. The most thorough investigators have repeatedly emphasized that in spite of apparent floristic homogeneity, the mycoflora varied considerably within short distances. Sappa (l.c.) systematically examined three apparently homogeneous sites of a Callunetum, some 60 m apart, for both vegetation and microfungi. The total number of plant species was 30, and 27 species occurred at each of the three sites. Fungal species totalled 96, but only 33 were present on all three sites, some with quantitative differences. Massari (1988) suggested to take soil samples at the corners of several concentric squares, in order to get a picture representative of the surface included, when the number of squares sampled in this way is correlated with the increasing numbers of species recorded, the surface required to obtain a representative picture of the species to be isolated can be estimated. But the idea of a sampling square, derived from macrophyte analysis, does not seem appropriate for microbiological sampling, as each subsample is representative only of one point, and the arrangement of the replicated sampling points is of little importance to the outcome.

The soil samples must be taken from a defined depth (soil horizon), either by removing all soil above the layer to be sampled or by sampling from the sides of a pit.

Care must be taken on the relation of the soil sample to plant roots. The *rhizosphere* phenomenon, an increased microbial density around living roots, has often been investigated since Hiltner (1904) coined the term. Depending on the purpose of the study, a sample is either taken at a distance from all plant roots or it will include roots. In the latter case, the roots are subsequently separated from the bulk and the adhering rhizosphere soil is afterwards removed.

In a mixed vegetation each plant species may selectively stimulate some fungal species in its rhizosphere (e.g. Summerbell 1989). Most analyses of the soil fungal flora have, however, been confined to models with a single or very few species of higher plants. Otherwise the pattern of distribution becomes too complex for an analysis. It is important to be aware of a strong variation of the microflora at the scale of *microhabitats* outlined here, when deciding on the sampling procedure for a particular purpose. Corresponding to the mosaic of microhabitats and the discontinuity of substrates, Swift (1984) and Swift and Heal (1986) speak of *unit communities* that occupy each 'resource unit'. Some analogies to island ecology can be demonstrated.

Soil samples must be processed as soon as possible after sampling. Any storage is bound to alter the microbial populations.

7.2.2. *Direct observation techniques*

Direct observation techniques have shown that many microfungi require soil cavities for sporulation (Kubiena and Renn 1935), but rarely allow the specific identification of the fungi involved. Most fungal hyphae observed directly are, however, sterile and often over 50% belong to basidiomycetes (Frankland 1982), i.e. usually macrofungi. Therefore, in contrast with the isolation techniques, this technique does not allow macrofungi to be distinguished from microfungi.

For this very selective review it suffices to state that the most successful direct observation technique so far is that of examining soil suspensions either mounted quantitatively in an agar film (Jones and Mollison 1948, with modifications by Thomas *et al.* 1965, Frankland 1975, Schnürer *et al.* 1985) or collected on membrane filters (Hanssen *et al.* 1974; simplified by Sundman and Sievelä 1978; reproducibility shown by Elmholt and Kjøller 1987) in order to quantify fungal biomass (Kjøller and Struwe 1982); the use of phase contrast microscopy (Frankland 1974) or staining with a fluorescent brightener (West 1988) further contribute to improve the results. The volume and weight of fungi per unit of soil can be calculated from length and diameter of the hyphae present, measurements of hyphal diameter being especially important (Bååth and Söderström 1979). Biomass estimates thus obtained for different soils range from less than 1 (peat, deep humus layers) to over 100 g m^{-2} (dry weight mycelium) in various forest and grassland soils (Kjøller and Struwe 1982).

Only a fraction of the mycelium observed is usually alive. Frankland (1975) tried to estimate the proportion of hyphae with live contents under phase contrast and got figures between 15 and 27% of the total fungal biomass. Staining with fluorescein diacetate (Söderström 1977), which fluoresces only after enzymatic hydrolysis, is theoretically the most elegant technique to assess the living component under the fluorescence microscope, but probably it leads to a considerable underestimation (Söderström 1979). Aceto-orcein has been used by Flanagan (1981) to make cytoplasm and nuclei visible; a further improvement is brought about by the use of europium chelate in combination with a fluorescent brightener (Johnen 1978). Such direct assessments of the fungal biomass may be more or less correlated with the activity of the total soil microflora, as measured by other, physiological techniques based on respirometry (Domsch *et al.* 1979, Schnürer *et al.* 1985) or determination of ergosterol content (Newell *et al.* 1988).

Several techniques have been devised, where a new substratum is buried for some time in the soil to allow fungal colonization. Unfortunately, the action of burying is bound to disturb the situation and to activate numerous micro-organisms; moreover, the newly introduced substratum provides an enrichment disturbance (see 7.3.3) selective for ruderally selected fungi. Waid and Woodman (1967) buried chemically inert nylon mesh to assess the production of fungal mycelium. This technique has rarely been used (Nagel-de Boois and Jansen 1971, Kuyper 1989). The dynamics of the mycelium present in the soil (production minus decomposition) can hardly be followed reliably even with this technique, as decomposition of mycelium starts very early while colonization is still in progress. As far as cellulolytic fungi are representative of soil fungal activity as a whole, Tribe's (1957) technique of burying cellophane squares supported by cover slips gives instructive pictures of fungal development in the soil. Gams (1960) modified the technique by fastening cellophane to a microscope slide and cutting 1 mm wide strips lengthwise before burying the slides, to facilitate the subsequent isolation of the fungi observed.

The connection between direct observation and qualitative analysis has been achieved only at the autecological level for certain species of fungi by labelling the hyphae of a particular species, e.g. with fluorescent antibodies (Frankland *et al.* 1981).

7.2.3. *Isolation techniques*

The most commonly used technique for isolation is the *dilution plate technique* (for description see, e.g., Johnson and Curl 1972); it has the advantages of easy use, yielding pure cultures mostly in one step, apparently quantitative results (assessment of numbers of colony-forming units = CFU per gramme soil), and is suitable for statistical analysis; but it has many drawbacks, the greatest being that usually more than 90% of the isolates obtained originate from resting stages (Warcup 1955b); strongly sporulating fungi are overestimated, whilst poorly sporulating or slow-growing fungi tend to be neglected or underestimated. The CFU is usually considered as the individual fungus, though the technique tends to disrupt the cells of what originally may have been one mycelium, sporangium, fruiting-body or conidiophore. The quantitative results often mislead mycologists into assessing the total number of fungi isolated from a particular soil as a measure of fungal density. This total count is completely meaningless. A quantitative count obtained with dilution plates is only meaningful when correlated with particular species. It is not possible with the technique to discriminate between active hyphae and inactive resting propagules. Nevertheless, the mere numerical information can be very important when certain species, particularly pathogenic fungi, are to be assessed.

Numerous modifications of the dilution plate technique have been devised either to improve homogenization or to adjust the isolation medium, i.e. to add compounds that might allow the isolation of some recalcitrant fungi. Detergents can be added to restrict lateral spread of fast-growing fungi, while the use of Rose Bengal for the same purpose is less to be recommended (it tends to suppress a considerable portion of the fungal spectrum entirely) (e.g. Gams and Van Laar 1982). These amendments had little success in extending the spectrum of fungi isolated, and an ordinary 2% malt agar is usually just as good for this purpose as more complex media.

Other modifications that render dilution plates and other techniques selective for a particular organism or group of organisms (e.g. cellulolytic fungi, Eggins and Pugh 1962, Park 1973) may increase their value considerably but only at the expense of its universality (surveys by Tsao 1970, Johnson and Curl 1972, for *Trichoderma* see Papavizas and Lumsden 1982).

The *Soil plate technique*, in which minute soil crumbs are dispersed in a Petri dish, was devised by Warcup (1950) to overcome some of the drawbacks of the dilution plates. It is a simple and very rapid technique that gives an impression of the variation in the soil fungal flora on a small scale. Only the percentage of Petri dishes containing a particular species is recorded; the spectrum of fungi isolated is the same as with dilution plates. One advantage of the dilution plates is that they give an overall/average picture of a larger soil volume.

Realizing that most of the isolates obtained with these techniques originated from resting stages (Warcup 1955a), Warcup (1955b) introduced a technique of isolating hyphae directly from the soil. A few workers who subsequently tried the method (including unpublished personal observations) had little success with it, because most hyphal fragments picked up from a washed soil sample turn out to be non-viable and efficient removal of all adhering spores cannot be guaranteed. Söderström and Erland (1986) refined the technique by selecting fluorescein diacetate-positive hyphae by micromanipulation.

Baiting fungi by means of living or dead plants, plant parts or other substrata is an important possibility for quantifying particular fungi that otherwise often escape analysis. But this has mainly been applied to plant-parasitic species.

The only real alternative to the isolation techniques listed above is *soil washing*. This technique was inspired by investigations with washed roots

(Harley and Waid 1955) and is aimed at isolating fungi adhering to the soil particles as hyphae. It was first applied to root-free soil by Parkinson and Williams (1961) and then modified repeatedly, a convenient simplified modification being that by Gams and Domsch (1967). A soil sample is agitated in water and fractionated through a range of screens of decreasing mesh width; from one or more of the finer screens, mineral and organic particles are then picked up after thorough washing and plated on agar. In Parkinson and Williams's technique the agitation is brought about by air bubbles blown through the system, in Gams and Domsch's by shaking the washing box (Figure 1). Inserting a vibrating disk ('Vibromischer') into a soil suspension is another suitable alternative for efficient washing (van Emden 1972).

For plating washed particles, media poor in nutrients, such as CMC agar (Gams 1960), soil extract or potato-carrot agar, are particularly suitable. On

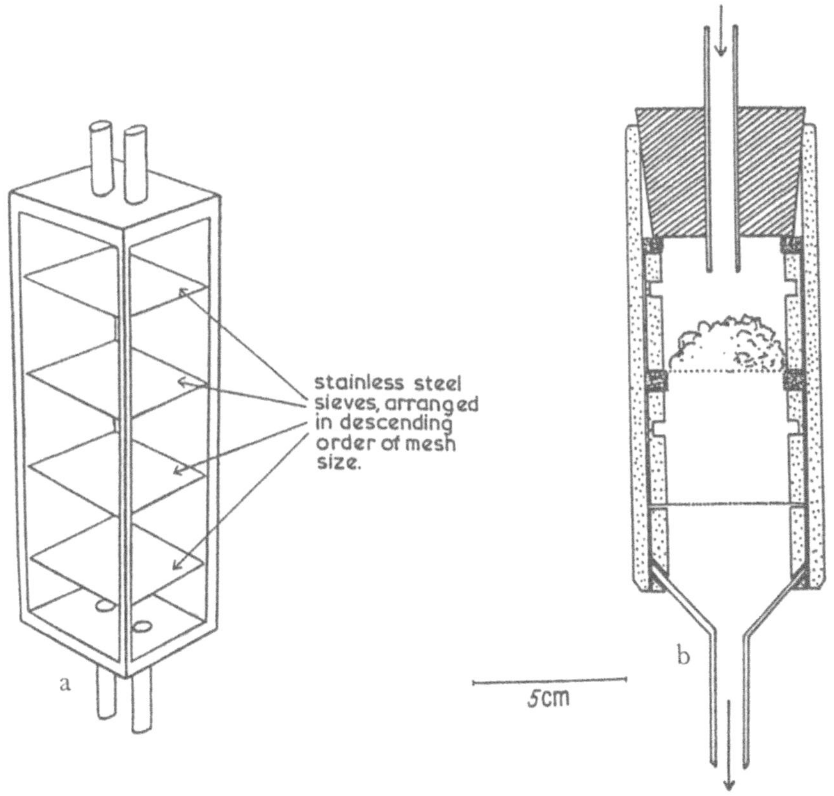

Fig. 1. Schematic representation of washing boxes as used by Parkinson and Williams (a) and Gams and Domsch (b).

these media fungi sporulate rapidly and can often be identified microscopically to species after about one week; moreover, when more than one species grow out from one particle, they can be separated only on weak media, whilst on acidified Czapek-Dox-yeast medium, originally used by Parkinson and Williams (1961) and other authors, the vigorous species tend to overgrow all the others. In comparative experiments using conifer rootlets (Schlag 1986, unpublished) it was found that the number of species observed on CMC agar was higher than on MEA and that no species were recovered on MEA that did not also appear on CMC. The soil washing technique requires a considerable amount of experience and often yields mixed cultures at the beginning. But it allows large numbers of isolates to be dealt with in a relatively short time. An additional advantage is the fact that aerial contaminants developing on the plates can be eliminated and all isolates obtained undoubtedly originate from the soil.

Calculation of the colonization quotient (CQ, the number of fungi per number of plated particles) gives a rough indication of hyphal density and washing efficiency. When mineral particles (sand grains) of about 1 mm diameter are plated, the CQ is usually below 1 and the chance of obtaining hyphal fragments after removal of all adhering spores is considered rather good, but with organic particles (root fragments etc., CQ values often between 2 and 3) this chance is much lower. Moreover, from organic particles, the proportion of fast-growing fungi (*Mucor, Mortierella, Fusarium* etc., Gams and Domsch 1969) is higher and that of slow-growing species lower, indicating that many of the latter tend to be overgrown and therefore overlooked.

The technique can be refined by extending the incubation period up to about two weeks (provided *Trichoderma* and other fast- and vigorously growing species are absent) to discover certain slow-growing fungi (Schlag 1986, unpublished), or by reducing the size of the organic particles to as little as 50 μm diameter (Bååth 1988), so that usually no more than one fungus can grow out from one particle. An additional advantage of the latter modification is that the data obtained when only one fungus grows out from each particle allow a better statistical analysis, as these data are independent, in contrast with those from the ordinary soil washing technique. Similar conclusions were reached by Kirby *et al.* (1990) who advocated the use of 0.04 mm^3 fragments of homogenized plant detritus from an aquatic biotope. These small particles can be compared to a sampling quadrat of plant vegetation analyses, where quadrat size is usually set so that only one species can have a frequency of 100%. Thus "the particle plating appears to be quite sensitive to changes in community composition in that it allows a great range of frequencies, probably because mycelium which is not involved in sporulation can be quantified" (Kirby *et al.* 1990).

When thousands of isolates have been obtained from one soil, the total spectrum of species obtained with the soil washing technique is almost identical with that of the dilution plates but quantitatively different (Gams and Domsch 1967). All the abundantly sporulating species dominating on dilution plates also appear on washed particles, provided care is taken to analyse all slow-growing species appearing amidst other fungi from one particle. On the other hand, it was surprising to find that very few additional species appeared on washed particles that had not been isolated by dilution plates, a notable exception being *Cladorrhinum foecundissimum*; this is a common species in agricultural soils, producing large numbers of phialoconidia that do not germinate, but also hyphal clusters which are ideally suited for isolating from washed particles (Gams and Domsch 1967, Domsch *et al.* 1980). The overall picture with washed soil particles is (a) a faster initial increase of species with increasing number of isolates (Gams and Domsch 1967), and (b) a quantitative shift towards species producing chlamydospores (Gams and Domsch 1969). The spectrum is apparently not shifted towards basidiomycetes. A surprising finding of our work was that the percentage of permanently sterile fungi was not increased among isolates obtained with the washing technique (about 3—4% of the isolates) when compared with dilution plates; in other words, it is possible with much patience to induce sporulation in a considerable proportion of the isolates, even after soil washing.

7.2.4. *Parameters for quantification and statistical analysis*

To quantify the proportion of individual components of fungal flora, various approaches have been pursued in view of an ecological interpretation. The crudest method is the indication of *numbers of species* present; the outcome often depends more on the intensity of the work and skill of the investigator than on the diversity of the population. A recording of presence or absence of a particular species in a sample depends to a large extent on its abundance. Most analyses are stopped when the curve relating the number of species with the number of isolates (see below) is still steep. Therefore, species with a low abundance have little chance of being detected.

With various isolation techniques it is possible to define the most important quantitative parameters for fungal analyses as follows: *abundance* (sometimes called density), the quantitative presence of a particular species within a single analysis, and *frequency*, the proportion with which a particular species is found among a number of temporally and/or spatially separated analyses (the latter term has, however, often been used for the former parameter). Thus, direct observation provides information of fungal abundance in general, soil dilution plates on the abundance of particular species.

The soil washing technique provides relative data of abundance, viz. the percentage of particles containing a particular species, while the washed particles from only a small and variable fraction of the total soil, different for each kind of soil. On the other hand, Warcup's soil plates yield only information on the spatial frequency of certain fungi within single samples. It is very difficult to compare and interpret studies in which different techniques have been used, unless only the presence or absence of species is taken into account (Domsch *et al.* 1980).

For the ecology of plant pathogens it is very important to assess not only the numerical frequency of particular organisms but also their *inoculum potential* (Garrett 1956). This parameter includes size of propagules, their nutritional state, etc. and can be assessed only at the autecological level. Lockwood (1988) has given a detailed account on the development of this and other soil-ecological concepts.

Thornton (1956) and subsequent workers noted the quantitative *dominance* of relatively few microfungi in different soils according to certain 'soil fungal patterns'. His technique of 'screened immersion plates' was claimed to recover only actively growing fungi. This is not necessarily true; it selects fast-growing fungi but does not recover any basidiomycetes. His observations again emphasize that the quantitative pattern of fungal distribution must be taken into account. This does not necessarily imply that only the numerically dominating species should be of ecological significance. Among the apparently rare species, some very important, e.g. pathogenic, species can be found, and the use of selective techniques is imperative for their study.

Statistical analysis. Whatever technique is used, extensive data sets of fungal analyses for soils differing in certain parameters must be analysed statistically to evaluate those parameters that really affect the fungal flora.

Statistical techniques allow the numbers of isolates that are needed to assess significant differences to be calculated, when the frequencies of some crucial species have been obtained.

The distribution of microfungi in soils does not usually obey the prerequisites of normal distribution and therefore either non-parametric tests, such as that by Wilcoxon and Wilcox (1964) can be used (Gams *et al.* 1969), or various transformations can be applied to the data before applying analyses of variance. Widden (1986a) found square root or negative binomal transformations to be suitable.

Analyses of soil fungal data have been done either for individual species or for the complete data sets, using methods of multivariate analysis, including discriminant analysis, factor or principal component analyses, multiple regression and canonical correlation analyses (Bissett and Parkinson 1979a–

c, Widden 1986a—c). An excellent introduction to these methods is given by Jongman *et al.* (1987).

The *diversity of the fungal population* can be assessed at two levels, (a) by simply recording the number of species recovered as a function of the number of isolates, and (b) by accounting for differences of abundance.

The number of species obtained during an analysis reflects the indigenous diversity only partly and the published fungal analyses rarely reach the flat part of the curve that relates fungal species with the number of isolates (Gams and Domsch 1967, Christensen 1981, Bååth 1981, see Figure 2).

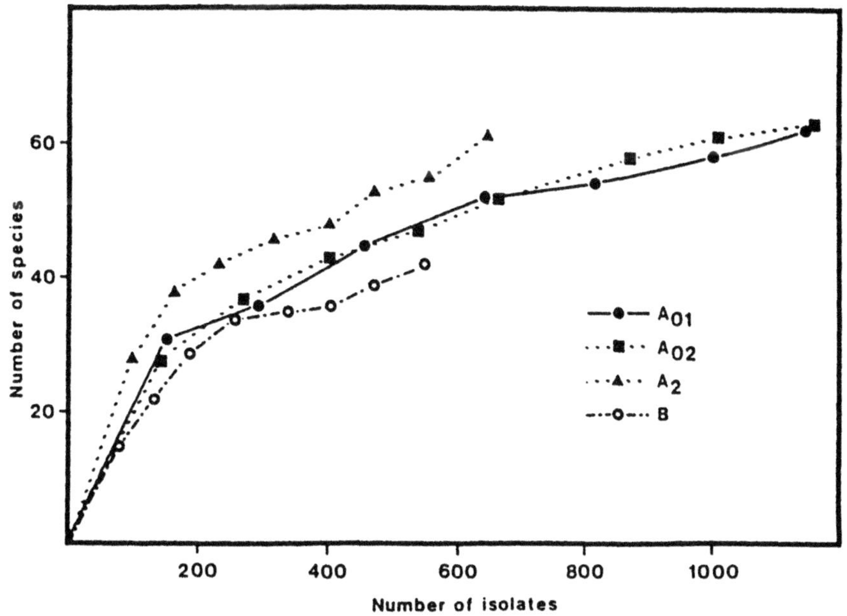

Fig. 2. The influence of the number of isolates on the number of species found in different horizons of a podzolic soil (from Bååth 1981).

Any published numbers of species should be seen in relation to the size of the total analysis to estimate the reliability of the data produced (see 7.3.4). Numbers of species also vary depending on the genera involved. In *Penicillium*, and some similar genera high numbers of species are more often recorded than in other, ecologically perhaps more important, genera. It also is relatively easy to differentiate *Penicillium* species, even macroscopically, and to estimate their numbers.

The similarity quotient according to Sørensen (1948),

$$S = \frac{2w}{a + b}$$

i.e. the number of common species divided by the species numbers in each list, has repeatedly been used for comparing species lists in different analyses, but, because of the above criticism, it is fraught with considerable inaccuracy.

A better estimate is obtained with the Bray and Curtis (1957) similarity index C,

$$C = \frac{2\Sigma w_i}{\Sigma(a_i + b_i)}$$

where the figures of abundance (a_i, b_i, the figures for the ith species in each population, w_i, the lower isolation frequency of that species in the two populations) enter into the calculation.

Christensen (1981) analysed 32 published reports using the Sørensen quotient in order to elucidate patterns of distribution for the dominant species; she also found some studies where microfungal diversity was correlated with vascular plant diversity.

A deeper knowledge of fungal communities from different biomes and from the horizons with each site is only possible when the relative importance of the identified fungi is known (Kjøller and Struwe 1982). For this purpose the Shannon-Weaver index H and an evenness index E are often used.

$H = -\Sigma\, p_i \log_2 p_i$, where p_i = importance probability of each species.

$$E = \frac{H}{(\log_2)\ \text{no. of species}}$$

The Shannon-Weaver function gives more pronounced numerical differences due to a greater sensibility for the number of species (more details in Morris and Rouse 1986). It was used to demonstrate a general trend of decreasing diversity with soil depth, though maximal values were observed in the F and H layers of forest soils (Kjøller and Struwe 1982).

It is interesting to record, that Van Emden (1972 and later, unpublished work), after many years of continuous studies, found a lower diversity in recently reclaimed Dutch polder soils, up to 14 years after reclamation, than in old land. But, contrary to expectation, the number of saprophytic fungi was not found to increase after several years of integrated or alternative agriculture when compared with conventional management (Verkerke 1978, and Nielander 1981, unpublished data).

7.2.5. *Conclusion on methodology*

Even the most universal techniques are more or less selective. Strongly sporulating species are over-represented, basidiomycetes and other groups under-represented.

To investigate a particular soil sample efficiently, it is recommended that several techniques are used concurrently. The final *choice of methods* depends on the purpose of the study. An overall quantitative estimation of fungal activity can be achieved only by direct observation and/or physiological methods.

Further criteria for a serious synecological inventory include: selection of homogeneous objects (for comparing the effects of certain parameters); adequate sampling with replications in space and time; extensive analyses with subsequent statistical analysis.

All isolation work must be guided by the necessity of a reliable *species identification.* Literature recommended for this purpose has been compiled by Gams *et al.* (1987). It is not sufficient to confine identifications to genus, because within a genus great variation of physiological and pathogenic capacities occurs. Checking of identifications by experts and permanent preservation of voucher cultures in culture collections is a good practice.

The limited amount of labour available usually necessitates a compromise between an adequate number of isolates from a single sampling and the required number of replicates in space and time required for a statistical analysis of the data.

7.3. The components of the soil mycoflora

The soil mycoflora comprises species of all major categories of fungi. Among the zoosporic fungi, most representatives require special baiting techniques not discussed here. Only the plant-pathogenic representatives of *Pythium* and *Phytophthora* are easily isolated on agar media, particularly when selective modifications are used (Flowers and Hendrix 1969, Tsao and Guy 1977, Van der Plaats-Niterink 1981). But the bulk of the soil fungi consists of Zygomycetes, Ascomycetes, Basidiomycetes and Deuteromycetes. Unfortunately, the ecologically important zygomycetous Endogonales (vesicular-arbuscular mycorrhizal fungi) escape all attempts of isolation.

Examples of species lists, two representative of agricultural and two of forest soils, are reproduced in Table I, from which the contrast in the spectrum between agricultural and forest soils, and possibly also climatic

Table I. The most frequently isolated species observed in representative analyses of two Dutch agricultural soils (abundance on washed organic particles, Wageningen and Oostelijk Flevoland, after Van Emden 1972) and their presence in some Swedish and Canadian coniferous forest soils (Söderström and Bååth 1978, Widden 1986a), with some adaptations of the nomenclature

	Wageningen	O-Flevo	Sweden	Canada
Plectosphaerella cucumerina		13.1		
Penicillium simplicissimum s. l.	8.2			+
Microdochium bolleyi		6.6		
Trichoderma hamatum	6.4			+
Mortierella alpina	5.6	3.9	+	
Mortierella elongata	5.6	5.3		
Paecilomyces carneus	5.1			+
Coniothyrium sp.	4.8	0.8		
Fusarium culmorum		4.2		
Phoma leveillei		2.9		
Cylindrocarpon destructans	4.3			+
Cladosporium cladosporioides		3.6	+	+
Zygorrhynchus moelleri	3.3			+
Penicillium janczewskii	2.7	0.8		
Verticillium nigrescens	2.7			
Fusarium equiseti		2.3		
Pseudeurotium zonatum	2.2			
Fusarium oxysporum	2.0			
Mortierella hyalina	1.9			
Gliocladium roseum	1.9			+
Trichoderma viride		37.5	+	+
Acremonium murorum		1.7		+
Phoma eupyrena		1.6		
Ramichloridium schulzeri		1.4		
Exophiala jeanselmei		1.3		
Volutella ciliata		0.9		
Trichoderma polysporum			+	+
Penicillium brevicompactum			+	+
Penicillium spinulosum			+	+
Penicillium thomii			+	+
Trichoderma koningii				+
Fusarium solani				+
Geomyces pannorum			+	+
Trichoderma harzianum		1.9		+

In addition the following species were frequent in the Swedish analysis: *Mortierella humilis, M. isabellina, M. macrocystis, M. verticillata, M. nana, M. parvispora, M. ramanniana, Mycelium radicis atrovirens, Oidiodendron* spp., *Tolypocladium inflatum*, and *Verticillium bulbillosum* (total number of species observed: 126).

regions, becomes evident. Surprisingly, among mesophilic heat-resistant fungi some more pronounced distributional patterns can be found (Table II).

From work of Frankland (1982) and others it is well known that basidiomycetes form a considerable proportion (often exceeding 50% of the fungal biomass) of the saprophytic soil fungal flora. Unfortunately no technique is so far available to isolate these basidiomycetes efficiently from the soil. Modest approaches in this direction are the hyphal isolation technique (Warcup 1955a), selective inhibitors in the agar media, viz. *o*-phenyl-phenol (Russell 1956) or benomyl (Maloy 1974, Stalpers 1976) but none of these modifications is effective in isolationg basidiomycetes from spores or single hyphae present in the soil, apart from some easily isolated species, such as *Coprinus* species, *Bjerkandera adusta* (Domsch and Gams 1970), and *Tephrocybe carbonaria* which is often isolated after a heat treatment (Bollen 1985, Table II). These have in common discrete small asexual (or sexual) propagules that can grow out on agar media while hyphal fragments usually lack this capacity.

7.3.1. *Taxonomic grouping*

A grouping according to major taxonomic categories is often adopted. Peyronel (1955) devised a graphic representation in octogons ('mycographs')

Table II. Habitats of ten selected heat-resistant fungi in the Netherlands (G. J. Bollen pers. comm.)

Origin of soils	Glasshouse-veget. garden	Field and pasture	Woodland	Heather
Number of samples	71	62	119	39
Average pH/KCl	6.6 ± 0.6	5.5 ± 1.0	4.0 ± 1.2	3.4 ± 0.3
	Frequency of occurrence (%)			
Gilmaniella humicola	75	34	8	3
Talaromyces flavus	30	48	8	0
Eupenicillium brefeldianum	37	21	12	5
Eupenicillium lapidosum	34	34	22	36
Neosartorya fischeri	44	29	25	3
Tephrocybe carbonaria	8	6	12	3
Trichophaea abundans	6	23	66	72
Mortierella vinacea	0	3	4	56
Eleutherascus tuberculatus	0	3	4	28

with eight radii representing species numbers of 'Phycomycetes', Coelomycetes, Tuberculariaceae, Aspergilli, Penicillia, Dematiaceae, Ascomycetes and remaining Moniliaceae; the method has been used by few other mycologists since. Using data from numerous authors, some staggering differences have been worked out by States (1981), but it must be borne in mind, that taxonomic grouping does not at all reflect ecological features of the fungi concerned. The criterion of numbers of species has been criticized above.

7.3.2. *Nutritional and ecological grouping*

The most meaningful ecological grouping follows the lines of succession in the decomposition of various plant-derived substrates (see 7.5). Bååth and Söderström (1980b), studying 60 species isolated from forest soil, found that 85% were able to decompose protein, 58% xylan, 52% cellulose and 38% chitin. Flanagan (1981) tabulated the decomposing capacities for pectin, starch, xylan, cellulose, humic and gallic acids and wood, found in numerous saprophytic soil fungi, the first three capacities being most widely distributed. A calculation of the percentage of active species has limited ecological value if their abundance in soil is ignored. Among several isolates of the same species, the capacities to decompose pectin, xylan and carboxymethyl cellulose varied little (Domsch and Gams 1969). Shifts in the population caused by crop rotation hardly altered the overall decomposition capacity in a soil (Domsch and Gams 1968b).

Typical inhabitants of decaying wood are the soft rot fungi of the genera *Phialophora, Chloridium* and *Oidiodendron* (Duncan 1960, Courtois 1963), which occur relatively infrequently among soil fungal analyses; greater densities of such fungi have, however, been noticed on tree roots (Harley and Waid 1955). Other genera, such as *Chalara, Hormiactella, Septonema* and many basidiomycetes are almost exclusively found on decaying wood and not normally isolated from soil. The activities of these fungi in wood decomposition include the formation of phenol oxidases (Holubová-Jechová 1971), but their effect is insignificant compared to that of basidiomycetes.

In a review on the metabolic versatility of fungi, Wainwright (1988) emphasized various other capacities, which provide ecological advantage to many soil fungi, particularly the potential of *oligocarbotrophic growth*: this may involve autotrophic CO_2 fixation (e.g. in *Fusarium*) or anaplerotic CO_2 fixation (i.e. fixation when small amounts of an organic substrate are supplied), which is known in many fungi, chemolithoautotrophic energy acquisition by oxidation of S, N, Mn or Fe compounds (probably of limited importance), or chemolithoheterotrophic growth, which combines the efficient utilization of low amounts of carbon sources with energy gain from inorganic oxidations. Strict faculative anerobiosis is a widespread phenomenon, but anaerobic respiration yields less energy than aerobic respiration. In addition, a very wide range of substrates can be utilized by soil fungi, including volatiles. The role of fungi in mineral cycling to make N, S and P compounds available is insufficiently appreciated. N, S, P, Fe and Mn compounds can not only be oxidized by fungi (with little energy yield) but also reduced; fungal nitrification is limited; fungal denitrification mostly takes place anaerobically but also aerobically and is of limited importance. The release of oxalate by fungi leads to calcium accumulation, acids and chelators contribute to the solubilization of rocks. *Oligotrophic growth* is normal for fungi in soil, whilst in the rhizosphere it can be characterized as *'modicotrophic' growth. 'Food-base heterotrophy'*, where the fungus gets its carbon directly from a particular substratum, occurs in mycorrhiza and wood decay, besides plant-parasitic fungi. Oligocarbotrophic growth is to be distinguished from nutrient-limited growth, a situation commonly observed during lignin decomposition (for a theoretical account see Bosatta and Berendse 1984). These potentials provide a range of additional parameters to be considered in the interpretation of fungal analyses. Conditions for oligotrophic growth should be created when the rôles of fungi and microbial interactions in the soil are to be analysed.

In the next paragraphs only three out of several ecological groups that have been studied individually are further considered.

7.3.2.1. *Thermophilic fungi*

Thermophilic fungi will only grow at temperatures above 20 °C (Cooney and Emerson 1964) and are therefore not usually recovered from soils in the temperate zones unless high temperatures are used for incubation. At temperatures of 37 and 48 °C, Apinis (1963a, b) retrieved considerable numbers of thermophilic fungi from English alluvial soils down to depths of 10—15 cm and these fungi have since been shown to be cosmopolitan.

7.3.2.2. *Osmophilic or xerophilic fungi*

Osmophilic fungi grow optimally on media with reduced water activity (containing 10—40% sugar or 10—20% NaCl) and can thus be isolated selectively. Well-known representatives are *Wallemia sebi* and species of *Eurotium.* They are commonly found on plant litter, fruits and fruit products, particularly in warmer zones, but not normally in soil.

7.3.2.3. *Nematophagous fungi*

Nematophagous fungi (and also amoebae- and rotifer-capturing fungi) comprise a highly specialized group of Zygomycetes, Hyphomycetes and Basidiomycetes that either capture and subsequently penetrate free-living nematodes, or the conidia attach themselves to, or are ingested by, their prey and subsequently invade the body (Barron 1977, Gray 1983, 1984). The delimitation between the two groups is not sharp. Nematophagous fungi appear rarely in ordinary soil fungal analyses but need specialized techniques for their recovery (Barron 1982, Gray 1984). There have been numerous studies in which these organisms were recorded in view of a possible application for biological control (Mankau 1980, Tribe 1977, 1980).

7.3.3. *The chemical elimination of certain functional groups*

It is possible to isolate particular components of the mycoflora selectively, e.g. Oomycetes with benomyl or polyene antibiotics, *Fusarium* species with PCNB (Tsao 1970), *or Penicillium* species using dichloran (King *et al.* 1979, Hocking and Pitt 1980), and certain other components of the microflora can be eliminated chemically. Thus, because of its defined spectrum of action, benomyl, and other MBC-generating systemic fungicides, are often used to suppress the majority of imperfect fungi (unless resistance has developed) (Bollen and Fuchs 1970, Edgington *et al.* 1971). Its overall effect on the microflora was studied by Hofer *et al.* (1971), Ponchet and Tramier (1971), and Siegel (1975); but non active fungi are not affected by it although many species are sensitive *in vitro* (Kaastra-Höweler and Gams 1973). Metalaxyl and similar fungicides can eliminate the Oomycetes, pencycuron (Monceren) has a similar effect on *Rhizoctonia solani.* Anderson and Domsch (1975) tested a range of compounds in an attempt to suppress fungal respiration. Only actidione had a satisfactory effect. But this compound is also used for the selective isolation from soil of dermatophytes belonging to the Onygenales. Streptomycin was found to be overall the most convenient suppressor of bacterial activity.

7.3.4. *Growth patterns, strategies*

Vegetative and generative growth patterns of the common soil fungi have repeatedly been distinguished and compared with their behaviour in pure culture, as far as growth rate, mycelial longevity, propensity and intensity of sporulation, etc. are concerned (e.g. Burges 1958). The extremes are the r-strategists with short-lived mycelia and abundant sporulation as in *Penicillium, Mucor* etc., attacking easily available substrates, and the K-strategists with a slower growing, more persistent mycelium, that does not sporulate inside the soil and attacks more resistant substrates.

The plant ecologist Grime (1979) distinguished three major life styles, largely determined by the three environmental facets of stress — disturbance — competition, to which plants are supposed to answer with stress-tolerant, ruderal, and competitive (combative) adaptations, respectively. Pugh (1980) and Pugh and Boddy (1988) have tried to adopt this scheme for soil fungi. Cooke and Rayner (1984) expanded the concepts, listing major components under each category:

Stress: long-lasting effects of growth-limiting, mostly abiotic, factors, such as extreme temperatures, moisture regimes, and pH values, shortage or intractability of nutrients, anthropogenic effects of pollution and xenobiotics (in particular heavy metals and pesticides). Stress tolerance is usually correlated with K-strategies.

Disturbance: sudden effects of either a *destructive* kind, such as fire, eating by animals, various anthropogenic effects, or *enrichment*, such as a surplus of organic matter provided in litter, compost heaps, baits for selective isolation etc. Disturbance effects can be of widely differing scale. The reaction to disturbance is ruderal adaptation (r-strategies above), i.e. fungi that have effective means of dispersal, germination, rapid uptake of nutrients and rapid mycelial extension, all of which allow rapid primary resource capture. Such fungi are generally ephemeral, non-combative, only capable of utilizing easily assimilable resources, and have rapid and often total commitment to reproduction.

Competition: combative ability allows an organism a) to defend its primary resource, b) to capture new secondary substrates; a good enzyme equipment is inherent in this capacity. Tolerance to stress and disturbance is less developed.

Various limiting ecological parameters can usually be attributed to one of these three categories. They usually affect fungal growth in various combinations. Hence the three categories can be represented in a triangle (Cooke and Rayner 1984) with the three extremes situated in the corners. On theoretical grounds it is better to discern first the morphological/physiological characteristics of micro-organisms which have been variously selected and then to check how they fit into the distribution patterns. To understand the behaviour of most fungi, it usually suffices to distinguish only the first mentioned two strategies.

7.4. The mycoflora of some important ecological niches

7.4.1. *Successions*

Two major groups of *fungal succession* can be distinguished (Park 1968, Frankland 1981), the substrate and the seral successions. The former leads from complex organic residues to more or less complete mineralization; this is the kind of succession most commonly studied with regard to saprophytic fungi. Seral succession follows the development of a biotope. Examples are the sequence of mobile to fixed dunes (Brown 1958), accompanied by comparable successions among higher plants, or the correlation of the gradient from pioneer to climax vegetation with certain spectra of microfungi (Tresner *et al.* 1954). However, it does not seem advisable to call any species from the first group a pioneer fungus, because in other soils the same species may show quite different relationships. After reclamation of Dutch polder soils, the population of *Trichoderma* gradually increased for five years, but later declined due to its sensitivity to bacterial antagonism (Pugh and Van Emden 1969). Still Trichoderma cannot generally be considered a genus of pioneer fungi.

The capacity to decompose resistant plant cell wall substances has led to the establishment of a sequence from sugar fungi to cellulolytic, ligninolytic, and keratinophilic fungi, which has often been referred to as the 'nutritional model' (Burges 1958, Garrett 1951, 1963, Frankland 1981). But secondary sugar fungi have repeatedly been noted to profit from the activity of cellulolytic predecessors (e.g. Hudson 1975). Obviously sugar fungi occur throughout the process of decomposition and are by no means confined to the early stages. This could also be predicted when disturbance is properly taken into account (Swift 1987).

The fructifications of coprophilous fungi often appear in a well-defined succession but no such sequence occurs in the mycelial growth which is active over the whole period (Webster 1970); this phenomenon may also play a rôle in other cases where successions have been described. Nevertheless, in the course of decomposition the resource quality is changed and this accounts for different fungal spectra observed in subsequent soil horizons.

The colonization of pine litter in various stages of decomposition and different soil horizons has been described in detail by Kendrick and Burges (1962), and many subsequent authors have restudied the successions on this and other substrata (for reviews see Dickinson and Pugh 1974, most recent study by Kloidt 1989).

Factors determining the success of fungal colonization include the 'Competitive saprophytic ability' and the 'Inoculum potential', two well-known terms coined by Garrett (1956); the former in particular implies an adequate enzyme equipment of the species. The two parameters must be in a mutual balance to lead to success. Three mechanisms are considered as determining succession:

1. earlier species pave the way for their successors
2. later species can grow better on low-level resources
3. early occupants may be killed by disturbance (Frankland 1981).

A non-equilibrium situation can be maintained by the impact of frequent disturbances on fungal assemblages. Huston (1979) speaks of a 'dynamic competitive equilibrium' that accounts for species diversity.

A succession of K-strategists after r-strategists has been suggested but this must be viewed with caution. Swift (1984) suggested an alternative terminology, occupation and window strategies, adapted to particular condition phases of the resource. As a unit community usually improves the quality of its resource in a first phase and later degrades it, it is bound to end up in a stress condition; an occupation strategist occupies the resource already during early stages of decomposition and is able to respond to changes so as to maintain occupancy (i.e., adaptation to persistence). A window strategist is able to respond to specific changes in the condition of the resource so as to secure occupancy in a particular stage. Viewed in this way, the three strategies of Grime would only operate at a secondary level (Swift 1984).

7.4.2. *Phyllosphere*

The components of the leaf-surface microflora are analysed by direct observation and isolation techniques. The community structure was discussed by Morris and Rouse (1986). A limited spectrum of yeasts (*Sporobolomyces, Rhodotorula, Cryptococcus*) and Hyphomycetes (*Aureobasidium, Cladosporium, Alternaria, Epicoccum*) is commonly observed. If there is a specific microbial association to be found anywhere, it is in the phyllosphere. The most recent reviews on the topic are found in Fokkema and Van den Heuvel (1986).

Endophytes of symptomless green parts of plants are rather specific in certain plant families (Petrini 1986). Some are latent pathogens, others have beneficial effects on their host (Carroll 1986, 1988). It is surprising to find not only species known as colonizers of senescent leaves, but also specific coprophilic ascomycetes and wood-inhabiting Xylariaceae among the endophytes (Riesen and Sieber 1985, Petrini and Petrini 1986).

7.4.3. *Litter*

The microfloras developing in various kinds of litter have been reviewed by numerous authors (Dickinson and Pugh 1974). Kloidt (1989) extended her studies towards the earliest phase of freshly fallen *Fagus* litter, where the former endophytes become the first decomposers. Findings of Kjøller and Struwe (1980) suggest the occurrence of stronger seasonal shifts in decomposition capacities within the fungal spectrum in the rapidly disintegrating *Alnus* litter than in deeper horizons.

7.4.4. *Soil depth*

Different fungal spectra have often been observed when the horizons of a soil profile were compared (e.g. Widden and Parkinson 1973, Visser and Parkinson 1975); such differences were also found to be statistically significant (Söderström 1975). Many species tend to decrease with soil depth, but others habitually occur further down. Most of the Mucorales are usually confined to the uppermost layers, species of *Penicillium, Paecilomyces, Acremonium, Sagenomella* and *Oidiodendron* commonly occur at greater depths. A passive washing down of fungal spores through the soil has been observed in an artificial model but is unlikely to play a major role in the natural soil (Hepple 1960). Tolerance of CO_2 by certain species contributes to their distribution in deeper soil layers (Burges and Fenton 1953). The organic substrates in subsequent soil horizons are in different stages of decomposition and humification; thus knowledge of the decomposition capacity of the fungal species may contribute to understanding the distribution pattern, but Bååth and Söderström (1980b) did not find considerable shifts in this respect in the populations of various horizons of podzol soils.

7.4.5. *Roots, rhizosphere*

An increased microbial activity in the vicinity of roots can be ascribed to root exudates, sloughed senescent root cells and mucigel, which have been described as rhizodeposition. While much progress is recorded in overall quantification of these effects (most recent review by Curl and Truelove 1986), limited new data are available on details of the fungal flora. A selection of fast-growing species among the soil fungi for selective colonization of the rhizosphere has repeatedly been documented (e.g. data for cereal roots compiled by Gams 1967) and different fungal species may be stimulated by different plants. Rootlets with ectomycorrhiza also are surrounded by a 'mycorrhizosphere', but the fungal spectrum was hardly different from that surrounding non-mycorrhizal root bark (Summerbell 1989). This author found a similar fungal spectrum in the mycorrhizosphere of *Picea mariana* as in the rhizosphere of *Cornus canadensis*, with only two species consistently differing.

7.4.6. *Dung*

Coprophilous fungi are specifically adapted to the passage through warm-blooded, mostly herbivorous animals. They become active in a selected community before and immediately after deposition of the faecal droppings. The composition of this community is determined by the fungal contents of the original feed (Webster 1970, Wicklow 1981a). Direct observation of the fructifications developing in moist chambers is an efficient means of study. Highly competitive fungi, such as *Stilbella fimetaria* or *Coprinus heptemerus*, may suppress the development of others. While most studies have been done on wet-incubated faecal material, incubation at reduced moisture (Kuthubutheen and Webster 1986) can allow additional species to grow out.

7.5. Ecological factors affecting the soil fungal flora

7.5.1. *Physical and chemical factors*

Moisture and soil atmosphere — Next to soil organic matter, moisture is usually considered the primary factor determining fungal activity. *In vitro* most fungi have their optimum at 100% r.h., but in the soil distinct patterns occur, particularly in the range between (85—)95 and 100% r.h., i.e. (-21—)-7—0 MPa (-70—0 bars) (see Griffin 1972). The reactions of different species to variations in humidity were also analysed by Widden (1986b, c). A certain similarity between winter and extreme summer aspects of the fungal flora may be ascribed to limited moisture availability. A raised CO_2 content, particularly in deeper layers, does not seem to affect the soil fungal pattern; only exceptional species like *Rhizoctonia solani* are so sensitive to CO_2, that they are limited to the uppermost soil layers (Domsch *et al.* 1980).

Temperature — In the temperate and cold zones, considerable numbers of soil fungi are psychrophilic and only a few thermophilic. Adaptation to low temperatures is well-known for tundra fungi (Flanagan 1981). Temperature is usually considered less important for fungal activity than soil moisture, but Widden (1986b) found this parameter to outweigh moisture in some Canadian forests. A winter community of fungal species was described.

Soil types — Soil characteristics that strongly affect the soil fungal population include organic matter content and pH (see below). Other properties determining soil types apparently have much less impact. Loub (1960) produced long species lists for different soil types. For a discussion of the inherent problems see 7.6.

Among the Mucorales it is relatively easy to find species with indicator value, characteristic of particular forest soil types, as found by Johann (1932) and Trautmann *et al.* (1992).

Soil organic matter — Different humus types show very different rates of organic matter decomposition (Handley 1954). The organic matter content of a soil is well known to affect the activity of soil microbial populations (Insam and Domsch 1988). But in several analyses of soil fungi in relation to kinds of vegetation (Widden 1986a), crop plants (Van Emden 1972), fertilization, and seasonal effects (Widden 1986b), it was found that major differences occurred between different sites that could not be attributed to particular soil or humus types.

The most important factor that affects soil fungal distribution is that of soil stratification and associated differences in the structure of soil organic matter.

pH — The investigations by Warcup (1951a) of five natural grassland soils of widely differing pH indicate the sensitivities of certain species. Species of *Humicola* and *Botryotrichum* as well as *Mortierella alpina* and *M. elongata* are characteristic of near-neutral soils, whilst other species of *Mortierella, Oidiodendron,* and many *Penicillium* species are more acidiphilic (also Linnemann 1958). Trautmann *et al.* (1992) observed rapid changes in the Mucorales flora following liming of acidic soils. The differences observed by Bissett and Parkinson (1979c) in the mycofloras of some alpine tundra sites could partly be attributed to pH effects, but also to different nitrogen contents. On the whole, a remarkable tolerance to acidification is observed among most saprophytic soil fungi, particularly those in naturally acidic biotopes, with some species even being stimulated (Bååth *et al.* 1984), whilst fungal activity is more strongly diminished. Strong acidification activates Al^{+3} ions which are toxic to most organisms including fungi (Rosswall *et al.* 1986).

Nitrogen and fertilization — The natural K and NO_3^- contents of forest soils were found to be among the parameters that affect the fungal spectrum (Widden 1986c). Rather few effects on the soil fungal flora have been observed after mineral nitrogen fertilization. Guillemat and Montégut (1956), using dilution plates, found fertilization of agricultural soils with NPK and stable manure to stimulate some fungal species. The effects of NPK were less pronounced than those of manure. The overall effect on the fungal spectrum was very limited. It is difficult to assess how far the positive effects of mineral fertilization can be attributed to an increased activity of the plants.

Fertilization of forest soils with ammonium nitrate or urea was found to increase tree growth in Sweden, but fungal activity was decreased (Bååth *et al.* 1981, Söderström *et al.* 1983). A few unusual species were stimulated by application of either urea or urine (Lehmann 1976).

7.5.2. *Different crop plants*

Different crop plants have in rotation usually little influence on the fungal spectrum (Domsch *et al.* 1968); only after continuous cropping with one plant were several species considerably increased in numbers (Van Emden 1972): under potato (partly also under sugar beet) *Plectosphaerella cucumerina, Acremonium furcatum* and *Volutella ciliata*, under sugar beet also *Verticillium nigrescens*, under ryegrass and wheat *Fusarium culmorum, Microdochium bolleyi, Acremonium murorum, Paecilomyces carneus* and *Phoma leveillei*. It is remarkable that some of the species listed did not show an increased abundance in the rhizosphere. In this study the effects of different locations again outweighed those of the crop plants. Plant-pathogenic fungi, however, may show a considerable accumulation in the rhizosphere.

7.5.3. *Seasonal variation*

As shown above, moisture and temperature affect fungal activity considerably and if they vary in a regular pattern, seasonal fluctuations of the soil mycoflora can be expected. Such effects have mostly been assessed only at generic level or above, and they are less conspicuous at the species level. Divergent trends of seasonal variation are found depending on the geographic region. Widden (1981) compiled studies on seasonal effects indicating the occurrence of spring and autumn, more rarely summer, maxima of activity. In spite of great temporal variations, Gams and Domsch (1969) found only a few species that showed any consistent seasonal effects in German soils.

7.5.4. *Disturbance and stress* (see also above, 7.3.4)

Major disturbances that affect the fungal flora are brought about either by forest fire (Widden and Parkinson 1975, Bissett and Parkinson 1980) or clear-cutting and removal of slash (Bååth 1981). After a fire, or also soil steaming (Warcup 1951b), a selection of heat-resistant fungi is observed, the propagules of which may be stimulated to germinate; another factor is recolonization by fast-growing fungi amongst which *Trichoderma* species predominate. For the fungi stimulated after fire, the term phoenicoid fungi has been proposed by Carpenter and Trappe (1985).

Acidification (artificial acid rain) was found to decrease the microbial biomass, and liming had a slightly less detrimental influence; a major shift occurred in the soil mesofauna (Bååth *et al.* 1980), but no data on fungal species are available. Pollution with heavy metals has repeatedly been shown to affect the soil fungal flora: along a gradient established by decreasing distance from a brass mill, Cu and Zn pollution was found to stimulate the occurrence of *Paecilomyces farinosus, Geomyces pannorum* and *Chalara* species, while some other species were adversely affected (Nordgren *et al.* 1985). An increased occurrence of *P. farinosus* may reflect a deleterious effect of heavy metals on insects; *G. pannorum* is well known for its tolerance to mercury and other toxicants.

The disturbance caused by ploughing agricultural soils is now often avoided by methods of reduced tillage intended to allow the development of a microbial equilibrium in the soil, a better preservation of organic matter in the uppermost layer and accompanying microbial stratification. The possible reactions to these and other agricultural measures have been commented on by Domsch (1986).

7.5.5. *Effects of the soil fauna*

The major role of earthworms in pedogenesis is well-known. The mor-type humus has few earthworms and greatest fungal densities (Handley 1954), but earthworms contribute positively to fungal development by substrate distribution and spread of propagules through faecal deposits. The main action of the micro- and mesofauna consists in grazing, and the overall effect on the fungal population can be either positive (stimulation of growth at low grazing rates) or negative (see contributions in Edwards *et al.* 1988).

Only a few studies have recorded fungi and other soil organisms simultaneously (e.g. Persson *et al.* 1980, Bååth *et al.* 1980). Microcosm models are particularly suitable for the analysis of the external factors affecting these interactions (Taylor and Parkinson 1988, Verhoef *et al.* 1988).

7.5.6. *Antagonistic effects of other micro-organisms*

Antagonistic actions of bacteria and streptomycetes, more than those of fungi, have been studied very frequently in respect of plant-pathogenic fungi (Fravel 1988). In recent years the often dramatic effects of siderophore-producing *Pseudomonas* strains have received most attention (Weller 1988).

Within a soil microhabitat, the fungal components can interfere with each other in various ways. There is a vast literature on dual interactions between microfungi *in vitro*. The phenomenon of antibiosis is, however, particularly difficult to demonstrate in soil (Brian 1957), and species that are incompatible *in vitro* can be found side-by-side on one soil particle (Gams and Domsch 1969, Bååth 1988). Thus it seems that the overall effect of fungus – fungus interactions in soil is much less than was anticipated from *in vitro* work, but combative interactions during the colonization of a particular substratum are bound to affect the developing population (Cooke and Rayner 1984).

A positive correlation of *Mortierella isabellina* with either *Penicillium spinulosum* or *P.* cf. *brevicompactum* on washed organic particles was noted by Bååth (1988). The kind of substratum might be responsible for such phenomena. Negative correlations among species of *Mortierella* could hardly be ascribed to antagonism but rather to the impossibility of recognizing the slow-growing species among colonies of the faster one.

The study of microbial interactions has benefitted from the advent of antifungal agents which helped to elucidate several direct and indirect effects of interactions and their consequences on plant disease (Bollen 1979).

Mycoparasitic interactions are bound to exist in the soil as well as *in vitro*. For example, growth of the ecologically obligate parasite *Verticillium biguttatum* can follow its host, *Rhizoctonia solani* along potato roots and stolons (Van den Boogert 1989).

7.6. Are there associations of microfungi in soils?

Very different kinds of organismal communities exist (Gee and Giller 1987). Soil fungi form rather loose and rapidly changing communities. But is there anything like a stable association characteristic of a biotope?

The species of microfungi are densely intermingled upon the substrata available in the soil. r-Strategists are apparently selected where extra nutrients are available, particularly around roots. For the rest it is very difficult to establish any regular patterns in those associations that can be correlated with units of vegetation or soil types.

Differences between fungal spectra of soils are of a quantitative kind and not qualitative. As a consequence, large numbers of isolates must be examined in each case to establish the absence of the less common species, in a particular soil with any degree of confidence.

Several investigators have tried to analyse microfungal associations in correlation with entites of the vegetation (e.g. Tresner *et al.* 1954, Apinis 1958, 1972, Christensen 1981, Widden 1986a). Fungal lists from different soil types have been worked out, e.g. by Loub (1960). Gams and Parkinson (1961) suggested that each plant species within a vegetation comprises a mosaic of different microhabitats for fungi (see 7.2.1) and the establishment of correlations between fungal spectra and the ecological factors determining vascular plant communities does not appear to be meaningful, but the type of vegetation rather than particular plant associations may well determine the soil microflora. An apparent correlation with soil types can also be related either to different climatic conditions, or to effects of differently structured soil horizons (see 7.4.4).

The fundamental difference between vegetation of plants and associations of microfungi is one of scale. Associations of higher plants are relatively stable; the microbial composition, on the other hand, changes rapidly in space and time (Burges 1970). In the soil a dense mosaic of microhabitats exists. One rain shower can alter the picture of the microflora drastically within a few days. Any attempt to establish communities of species in a particular biotope presupposes the analysis of composite samples representative of a larger area of an apparently homogeneous biotope; thus the variation at micro-scale is artificially smoothed out and only the mass of numerically dominating, sometimes locally separated, species becomes apparent.

Soil fungal decomposer communities have been summed up by Swift and Heal (1986) and Swift (1987). There is essentially a community mosaic. "Within a given vegetation type, the fungi are partitioned between the different types and species of resource. The fungi of leaf litter are substantially (though not entirely) different to those of branches, roots or faecal pellets (. . .) A distinction can be drawn between resource specific fungi and resource non-specific species (. . .) The limits and character of such specificity in the decomposer fungi have, however, not been well-defined and warrant further investigation." The fungal mosaic operates within a larger mosaic determined by the meso- and macro-scale activities of the fauna.

As stated above, soil fungi are subject to soil mycostasis and remain in a dormant state for most of their life. But, in addition, soil-borne species found in a soil A can hardly establish themselves in another soil B where they had previously been absent. Apparently the equilibrium built up in each soil has taken a long time to develop.

Many common species appear to be distributed erratically over widely differing biotopes. It is, for example, not at all understood, why particular species of the commonest genera, *Trichoderma* and *Penicillium*, occur in a soil A and are replaced by other species in a similar soil B. On the other hand, identical species occur in widely differing soils. Species that show statistically significant correlations with either coniferous or deciduous forest in one particular case (e.g. Widden 1986a) may be dominant in the other or in quite different biotopes elsewhere. While the ecological amplitude in relation to most environmental factors is very broad, the actual pattern of distribution is much more limited. Thus, in contrast with plants, the occurrence of which is mainly determined by known ecological factors (besides the overall area of the species involved), in the distribution of microfungi the stochastic element (advantage of the first occupant) seems to play an important rôle. The situation can be paraphrased by saying that their distribution depends on some ecological factors and, to a large extent, on chance factors that are not yet fully understood.

A notable feature is that the quantitatively dominant species have high communality values in the comparison of different biotopes, the infrequent ones show a large amount of unique variation. Therefore Bissett and Parkinson (1979b) concluded that the commonly isolated species 'must be dominant members of the community and therefore have a controlling influence on the development of other populations (. . .) species with low communalities were probably incidentals which were indifferent to the activities of the other species." But Widden (1986a) found some exceptions to this rule.

There are very few species amongst the commonest soil fungi to which an *indicator role* for particular ecological factors can be attributed. Besides a distinction of major climatic zones (e.g. Russian studies by Mishustin *et al.* 1961, Mishustin 1972), only very crude differences between biotopes, such as forest or heathland soils on the one hand, and agricultural and meadow soils on the other (see Table I), or very acidic vs. alkaline soils, temperate vs. tropical soils, are reflected by some of their fungal components (Christensen 1981, Domsch *et al.* 1980). A rare case of a species highly specific of heathland soils is the heat-resistant *Eleutherascus tuberculatus*, although the decisive factor of this phenomenon is unknown (Bollen and Van der Pol-Luiten 1975, and Table II). The patterns worked out by Christensen (1981) illustrate the impossibility of clear-cut distinctions for the commonest species.

7.7. Conclusion

Frankland (1981) asked the crucial question: "Can the mycologist advance beyond the compiling of species lists in the study of succession?" and found that few studies have done so. Nevertheless, considerable progress has been made in the last two decades in the study of the soil micromycete flora. For such studies to be ecologically meaningful, much patient work is required over a prolonged time. The requirements have been summarized under 7.2.6.

The effects of certain ecological factors on the mycoflora can only be assessed when the same investigator(s) study closely comparable objects, so that the role of single parameters can be evaluated (often after they have been varied artificially). When more widely differing samples are compared, the number of ecological factors involved is so high, that it is hardly possible to apportion the observed differences to a particular ecological factor.

Once the spectrum of species has been determined, the names disclose much information available in the literature (e.g. Domsch *et al.* 1980) on the ecological/physiological potential of these organisms. Although some conclusions on the ecological significance can be drawn at generic level (e.g. chitin decomposition in *Mortierella*, hemicellulose and cellulose decomposition by *Penicillium*, etc.), there are considerable differences at species level that make detailed identification necessary.

The enormous amount of work involved in such investigations must be related to the purpose of the study. There are many cases where the mere assessment of fungal biomass or the quantification of some components of the mycoflora, e.g. the cellulolytic fungi, or certain autecological models can satisfactorily answer a question. Such straight-forward approaches are of particular value when the principal aim is to understand a process rather than the population.

For many practical purposes it is adequate to study certain pathogenic or antagonistic soil-borne species autecologically. They live in the soil in equilibrium with saprophytic fungi, bacteria and the microfauna, and, according to recent findings (Edwards *et al.* 1988), the latter two components tend to outweigh fungal antagonists in their importance for the persistence of a single fungal species.

There is always a multiplicity of decomposer fungi which break down the organic substrates jointly. "This contributes to a stabilized functional profile within decomposer communities" (Swift and Heal 1986). It is in line with this statement that the saprophytic microfungal flora of the soil is remarkably little affected by various agricultural practices, including fertilization, certain crop plants and plant protection chemicals. Their effects are stronger in the rhizosphere than in the root-free soil. Other ecological parameters considered singly often affect only some of the species quantitatively.

Acknowledgement

I am greatly indebted to Drs E. Arnolds, G. J. Bollen, Th. W. Kuyper, B. E. Söderström and W. Winterhoff for suggesting improvements on drafts of this paper. Dr Sheila M. Francis meticulously edited it linguistically.

References

Anderson, J. P. E., and K. H. Domsch. 1975. Measurement of bacterial and fungal contributions to respiration of selected agricultural and forest soils. Can. J. Microbiol. 21: 314—322.

Anderson, J. P. E., and K. H. Domsch. 1978a. Mineralization of bacteria and fungi in chloroform-fumigated sils. Soil Biol. Biochem. 10: 207—213.

Anderson, J. P. E., and K. H. Domsch. 1978b. A physiological method for the quantitative measurement of microbial biomass in soils. Soil Biol. Biochem. 10: 215—222.

Apinis, A. E. 1958. Distribution of microfungi in soil profiles of certain alluvial grasslands. Angew. Pflanzensoziol. 15: 83—90.

Apinis, A. E. 1963a. Thermophilous fungi of coastal grasslands. p. 427—438. *In*: J. Doeksen and J. van der Drift (eds.) Soil Organisms. North-Holland Publ. Co., Amsterdam.

Apinis, A. E. 1963b. Occurrence of thermophilous microfungi in certain alluvial soils near Nottingham. Nova Hedwigia 5: 57—78.

Apinis, A. E. 1972. Facts and problems. Mycopath. Mycol. appl. 48: 93—109.

Baas-Becking, L. G. M. 1934. Geobiologie of inleiding tot de milieukunde. Van Stockum, Den Haag.

Bååth, E. 1981. Microfungi in a clear-cut pine (*Pinus sylvestris*) forest soil in central Sweden. Can. J. Bot. 59: 1331—1337.

Bååth, E. 1988. A critical examination of the soil washing technique with special reference to the effect of the size of the soil particles. Can. J. Bot. 66: 1566—1569.

Bååth, E., B. Berg, U. Lohm, B. Lundgren, H. Lundkvist, T. Rosswall, B. E. Söderström, and A. Wirén. 1980. Effects of experimental acidification and liming on soil organisms and decomposition in a Scots pine forest. Pedobiologia 20: 85—100.

Bååth, E., B. Lundgren, and B. E. Söderström. 1981. Effects of nitrogen fertilization on the activity and biomass of fungi and bacteria in a podzolic soil. Zentbl. Bakt. Hyg., Abt. 1, Orig. C 2: 90—98.

Bååth, E., B. Lundgren, and B. E. Söderström. 1984. Fungal populations in podzolic soil experimentally acidified to simulate acid rain. Microbial Ecol. 10: 197—203.

Bååth, E., and B. E. Söderström. 1978. Microfungi in Swedish coniferous forest soils. Svensk bot. Tidskr. 72: 343—349.

Bååth, E., and B. E. Söderström. 1979. The significance of hyphal diameter in calculation of fungal biovolume. Oikos 33: 11—14. 1979.

Bååth, E., and B. E. Söderström. 1980a. Comparisons of the agar-film and membrane-filter methods for the estimation of hyphal lengths in soil, with particular reference to the effect of magnification. Soil Biol. Biochem. 12: 385—387.

Bååth, E., and B. E. Söderström. 1980b. Degradation of macromolecules by microfungi isolated from different podzolic soil horizons. Can. J. Bot. 58: 422—425.
Barron, G. L. 1977. The nematode-destroying fungi. Can. Biol. Publ., Guelph, Ontario.
Barron, G. L. 1982. Nematode-destroying fungi. p. 533—552. *In*: R. G. Burns and J. H. Slater (eds.) Experimental Microbiology. Blackwell, Oxford.
Bissett, J., and D. Parkinson. 1979a. The distribution of fungi in some alpine soils. Can. J. Bot. 57: 1609—1629.
Bissett, J., and D. Parkinson. 1979b. Fungal community structure in some alpine soils. Can. J. Bot. 57: 1630—1641.
Bissett, J., and D. Parkinson. 1979c. Functional relationships between soil fungi and environment in alpine tundra. Can. J. Bot. 57: 1642—1659.
Bissett, J., and D. Parkinson. 1980. Long-term effects of fire on the composition and activity of the soil microflora of a subalpine, coniferous forest. Can. J. Bot. 58: 1704—1721.
Boerema, G. H. and L. H., Höweler. 1967. *Phoma exigua* and its varieties. Persoonia 5: 15—28.
Bollen, G. J. 1979. Side-effects of pesticides on microbial interactions. p. 451—481. *In*: B. Schippers and W. Gams (eds.) Soil-borne plant pathogens. Academic Press, London.
Bollen, G. J. 1985. Lethal temperatures of soil fungi. p. 191—193. *In*: C. A. Parker *et al.* (eds.) Ecology and management of soilborne plant pathogens. Am. Phytopath. Soc., St. Paul.
Bollen, G. J., and A. Fuchs. 1970. On the specificity of the in vitro and in vivo antifungal activity of benomyl. Neth. J. Pl. Path. 76: 299—312.
Bollen, G. J., and B. van der Pol-Luiten. 1975. Mesophilic heat-resistant fungi. Acta bot. neerl. 24: 254—255.
Boogert, P. H. J. F. van den. 1989. Colonization of roots and stolons of potato by the mycoparasitic fungus *Verticillium biguttatum*. Soil Biol. Biochem. 21: 255—262.
Bosatta, E., and F. Berendse. 1984. Energy or nutrient regulation of decomposition: implications for the mineralization-immobilization response to perturbations. Soil Biol. Biochem. 16: 63—67.
Bray, J. R., and J. T. Curtis. 1957. An ordination of the upland forest communities of southern Wisconsin. Ecol. Monogr. 27: 325—349.
Brian, P. W. 1957. The ecological significance of antibiotic production. p. 168—188. *In*: R. E. O. Williams and C. C. Spicer (eds.) Microbial ecology. 7th Symp. Soc. gen. Microbiol.
Brown, J. C. 1958. Soil fungi of some British sand dunes in relation to soil type and succession. J. Ecol. 46: 641—664.
Burges, N. A. 1958. Micro-organisms in the soil. Hutchinson University Library, London.
Burges, N. A. 1963. The microbiology of a podzol profile. p. 151—157. *In*: J. Doeksen and J. van der Drift (eds.) Soil organisms. North Holland Publ. Co., Amsterdam.
Burges, N. A. 1970. Time and size as factors in ecology. J. Ecol. 48: 273—285.
Burges, N. A. and E. Fenton. 1953. The effect of carbon dioxide on the growth of certain soil fungi. Trans. Br. mycol. Soc. 36: 104—108.
Carpenter, S. E., and J. M. Trappe. 1985. Phoenicoid fungi: a proposed term for fungi that fruit after heat treatment of substrates. Mycotaxon 23: 203—206.
Carroll, G. C. 1986. The biology of endophytism in plants, with particular reference to woody perennials. p. 205—222. *In*: N. J. Fokkema and J. van den Heuvel (eds.) Microbiology of the phyllosphere. Cambridge Univ. Press.
Carroll, G. C. 1988. Fungal endophytes in stems and leaves: from latent pathogens to mutualistic symbiont. Ecology 69: 2—9.
Christensen, M. 1981. Species diversity and dominance in fungal communities. p. 201—232. *In*: D. T. Wicklow and G. C. Carroll (eds.) The fungal community, its organization and role in the ecosystem. M. Dekker, New York, Basel.
Christensen, M. 1989. A view of fungal ecology. Mycologia 81: 1—19.
Cooke, R. C., and A. D. M. Rayner. 1984. Ecology of saprotrophic fungi. Longman, London.

Cooke, R. C., and J. M. Whipps. 1980. Evolution of modes of nutrition in fungi parasitic on terrestrial plants. Biol. Revs. 55: 341—362.

Cooney, D. G., and R. Emerson. 1964. Thermophilic fungi. Freeman, S. Francisco and London.

Courtois, H. 1963. Beitrag zur Frage holzabbauender Ascomyceten und Fungi imperfecti. Holzforschung 17: 176—183.

Curl, E. A., and B. Truelove. 1986. The rhizosphere. Springer, Berlin, Heidelberg.

Dickinson, C. H., and G. J. F. Pugh (eds.). 1974. Biology of plant litter decomposition. Academic Press, London.

Dobbs, C. G., and W. H. Hinson. 1953. A widespread fungistasis in soils. Nature, Lond. 172: 197—199.

Domsch, K. H. 1986. Influence of management on microbial communities in soil. p. 355—367. *In*: V. Jensen, A. Kjøller and L. H. Sørensen (eds.) Microbial communities in soil. Elsevier, London.

Domsch, K. H., Th. Beck, J. P. E. Anderson, B. Söderström, D. Parkinson, and G. Trolldenier. 1979. A comparison of methods for soil microbial population and biomass studies. Z. Pflanzenern. Bodenk. 142: 520—533.

Domsch, K. H., and W. Gams. 1968a. Die Bedeutung vorfruchtabhängier Verschiebungen in der Bodenmikroflora. I. Der Einfluß von Bodenpilzen auf die Wurzelentwicklung von Weizen, Erbsen und Raps. Phytopath. Z. 63: 64—74.

Domsch, K. H., and W. Gams, 1968b. Die Bedeutung vorfruchtabhängier Verschiebungen in der Bodenmikroflora. III. Der Abbau organischer Substrate. Phytopath. Z. 63: 287—297.

Domsch, K. H., and W. Gams. 1969. Variability and potential of a soil fungus population to decompose pectin, xylan and carboxymethyl-cellulose. Soil Biol. Biochem. 1: 29—36.

Domsch, K. H., and W. Gams. 1970. Pilze aus Agrarböden. G. Fischer, Stuttgart.

Domsch, K. H., W. Gams, and T.-H. Anderson. 1980. Compendium of Soil Fungi. Academic Press, London.

Domsch, K. H., W. Gams, and E. Weber. 1968. Der Einfluß verschiedener Vorfrüchte auf das Bodenpilzspektrum in Weizenfeldern. Z. PflErn. Bodenk. 119: 134—149.

Duncan, C. G. 1960. Wood-attacking capacities and physiology of soft-rot fungi. Rep. U.S. Dep. Agric., For Prod. Lab. Madison No. 2173.

Edington, L. V., K. L. Khew, and G. L. Barron. 1971. Fungitoxic spectrum of benzimidazole compounds. Phytopathology 61: 42—44.

Edwards, C. A., B. R. Stinner, D. Stinner, and S. Rabatin (eds.). 1988. Biological interactions in soil. Elsevier, Amsterdam.

Eggins, H. O. W., and G. J. F. Pugh. 1962. Isolation of cellulose-decomposing fungi from the soil. Nature, Lond. 207: 440—441.

Elmholt, S., and A. Kjøller. 1987. Measurement of the length of fungal hyphae by the membrane filter technique as a method for comparing fungal occurrence in cultivated field soils. Soil Biol. Biochem. 19: 679—682.

Emden, J. H. van, 1972. Soil mycoflora in relation to some crop-plants. EPPO Bull. 7: 17—26.

Flanagan, P. W. 1981. Fungal taxa, physiological groups, and biomass: a comparison between ecosystems. p. 509—592. *In*: D. T. Wicklow and G. C. Carroll (eds.) The fungal community, its organization and role in the ecosystem. M. Dekker, New York, Basel.

Flowers, R. A., and J. W. Hendrix. 1969. Gallic acid in a procedure for isolation of *Phytophthora parasitica* var. *nicotianae* and *Pythium* spp. from soil. Phytopathology 59: 725—731.

Fokkema, N. J., and J. van den Heuvel (eds.). 1986. Microbiology of the phyllosphere. Cambridge Univ. Press.

Frankland, J. C. 1974. Importance of phase-contrast microscopy for estimation of total fungal biomass by the agar-film technique. Soil Biol. Biochem. 6: 409—410.

Frankland, J. C. 1975. Estimation of live fungal biomass. Soil Biol. Biochem. 7: 339—340.

Frankland, J. C. 1981. Mechanisms in fungal succession. p. 403—426. *In*: D. T. Wicklow and

G. C. Carroll (eds.) The fungal community, its organization and role in the ecosystem. M. Dekker, New York, Basel.
Frankland, J. C. 1982. Biomass and nutrient cycling by decomposer Basidiomycetes. p. 241—261. *In*: J. C. Frankland, J. N. Hedger and M. J. Swift (eds.) Decomposer Basidiomycetes, their biology and ecology. Cambridge University Press.
Frankland, J. C. 1990. Ecological methods of observing and quantifying soil fungi. Trans. mycol. Soc. Japan 31: 89—101.
Frankland, J. C., A. D. Bailey, T. R. G. Gray, and A. A. Holland, 1981. Development of an immunological technique for estimating mycelial biomass of *Mycena galopus* in leaf litter. Soil Biol. Biochem. 13: 87—92.
Fravel, D. R. 1988. Role of antibiosis in the biocontrol of plant diseases. A. Rev. Phytopath. 26: 75—91.
Gams, W. 1960. Studium zellulolytischer Bodenpilze mit Hilfe der Zellophanstreifen-Methode und mit Carboxymethyl-Zellulose. Sydowia 14: 295—307.
Gams, W. 1967. Mikroorganismen in der Wurzelregion von Weizen. Mitt. Biol. Bundesanst. Ld.- u. Forstw. 123: 77 pp.
Gams, W., H. A. van der Aa, A. J. van der Plaats-Niterink, R. A. Samson, and J. A. Stalpers. 1987. CBS course of mycology. Centraalbureau voor Schimmelcultures, Baarn.
Gams, W., and K. H. Domsch. 1967. Beiträge zur Anwendung der Bodenwaschtechnik für die Isolierung von Bodenpilzen. Arch. Mikrobiol. 58: 134—144.
Gams, W., and K. H. Domsch. 1969. The spatial and seasonal distribution of microsopic fungi in arable soils. Trans. Br. mycol. Soc. 52: 301—308.
Gams, W., K. H. Domsch, and E. Weber. 1969. Nachweis signifikant verschiedener Pilzpopulationen bei gleicher Bodennutzung. Pl. Soil 31: 439—450.
Gams, W., and W. van van Laar. 1982. The use of Solacol ® (validamycin) as a growth retardant in the isolation of soil fungi. Neth. J. Pl. Path. 88: 39—45.
Gams, W., and D. Parkinson. 1961. Problematik der Bodenmykologie. Ber. geobot. Inst. Eidg. Techn. Hochsch. St. Rübel 32: 176—186.
Garrett, S. D. 1951. Ecological groups of soil fungi; a survey of substrate relationships. New Phytol. 50: 149—166.
Garrett, S. D. 1956. Biology of root-infecting fungi. Cambridge Univ. Press.
Garrett, S. D. 1963. Soil fungi and soil fertility. Pergamon Press, Elmsford, New York.
Gee, J. H. R., and P. S. Giller (eds.). 1987. Organization of communities, past and present. Blackwell, Oxford.
Gray, N. F. 1983. Ecology of nematophagous fungi: distribution and habitat. Ann. appl. Biol. 102: 501—509.
Gray, N. F. 1984. Ecology of nematophagous fungi: methods of collection, isolation and maintenance of predatory and endoparasitic fungi. Mycopathologia 86: 143—153.
Griffin, D. M. 1972. Ecology of soil fungi. Chapman and Hall, London.
Grime, J. P. 1979. Plant strategies and vegetation processes. J. Wiley, Chichester.
Guillemat, J., and J. Montégut. 1956. Contribution à l'étude de la microflore fongique des sols cultivés. Annls Épiphyt. 7: 472—540.
Handley, W. R. C. 1954. Mull and mor formation in relation to forest soils. Bull. For. Comm. London 23: 115 pp.
Hanssen, J. F., T. F. Thingstad, and J. Goksøyr. 1974. Evaluation of hyphal length and fungal biomass in soil by a membrane filter technique. Oikos 25: 102—107.
Harley, J. L., and J. S. Waid. 1955. A method of studying active mycelia on living roots and other surfaces in the soil. Trans. Br. mycol. Soc. 38: 104—118.
Hepple, S. 1960. The movement of fungal spores in soil. Trans. Br. mycol. Soc. 43: 73—79.
Hiltner, L. 1904. Über neuere Erfahrungen und Probleme auf dem Gebiet der Bodenbakteriologie und unter besonderer Berücksichtigung der Gründüngung und Brache. Arb. dt. Landw. Ges. 98: 59—78.

Hocking, A. D., and J. I. Pitt. 1980. Dichloran-glycerol medium for enumeration of xerophilic fungi from low-moisture foods. Appl. environm. Microbiol. 39: 488—492.
Hofer, I., Th. Beck, and P. Wallnöfer. 1971. Der Einfluß des Fungizids Benomyl auf die Bodenmikroflora. Z. PflKrankh. PflSchutz 78: 398—407.
Holubová-Jechová, V. 1971. Phenoloxidase enzymes from wood-inhabiting Hyphomycetes. Česká Mykol. 25: 23—32.
Hudson, H. J. 1975. Secondary saprophytic fungi. p. 15—18. *In*: G. Kilbertus *et al.* (eds.) Biodégradation et humification. Nancy. Pierron, Sarreguémines.
Huston, M. 1979. A general hypothesis of species diversity. Am. Natur. 113: 81—101.
Ingold, C. T. 1971. Fungal spores, their liberation and dispersal. Clarendon, Oxford.
Insam, H., and K. H. Domsch. 1988. Relationship between soil organic carbon and microbial biomass on chronosequences of reclamation sites. Microb. Ecol. 15: 177—188.
Johann, F. 1932. Untersuchungen über Mucorineen des Waldbodens. Zentbl. Bakt. ParasitKde, Abt. 2, 85: 305—338.
Johnen, B. G. 1978. Rhizosphere micro-organisms and roots stained with europium chelate and fluorescent brightener. Soil Biol. Biochem. 10: 495—502.
Johnson, L. F. and E. A. Curl. 1972. Methods for research on the ecology of soil-borne plant pathogens. Burgess Publ. Co., Minneapolis.
Jones, P. T. C., and J. E. Mollison. 1948. A technique for the quantitative estimation of microorganisms. J. gen. Microbiol. 2: 54—69.
Jongman, R. H. G., C. J. F. ter Braak, and O. F. R. van Tongeren. 1987. Data analysis in community and landscape ecology. PUDOC, Wageningen.
Kaastra-Höweler, L. H., and W. Gams. 1973. Preliminary study on the effect of benomyl on the fungal flora in a greenhouse soil. Neth. J. Pl. Path. 79: 156—158.
Kendrick, W. B., and N. A. Burges. 1962. Biological aspects of the decay of *Pinus sylvestris* leaf litter. Nova Hedwigia 4: 313—342.
King, A. D., A. D. Hocking, and J. I. Pitt. 1979. Dichloran-rose bengal medium for the enumeration and isolation of molds from foods. Appl. environm. Microbiol. 37: 959—964.
Kirby, J. J. H., J. Webster, and J. H. Baker. 1990. A particle plating method for analysis of fungal community composition and structure. Mycol. Res. 94: 621—626.
Kjøller, A., and S. Struwe. 1980. Microfungi of decomposing red alder leaves and their substrate utilization. Soil Biol. Biochem. 12: 425—431.
Kjøller, A., and S. Struwe. 1982. Microfungi in ecosystems: Fungal occurrence and activity in litter and soil. Oikos 39: 1391—1422.
Kloidt, M. 1989. Untersuchungen zum Abbau der Buchenblattstreu durch Pilze. Dissert. bot. 130, J. Cramer.
Kubiena, W., and C. E. Renn. 1935. Micropedological studies of the influence of different organic compounds upon the microflora of the soil. Zentbl. Bakt. ParasitKde, Abt. 2, 91: 267—292.
Kuthubutheen, A. J., and J. Webster. 1986. Water availability and the coprophilous fungus succession. Trans. Br. mycol. Soc. 86: 63—76.
Kuyper, Th. W. 1989. Mycelial growth in a pine forest subjected to different fertilisation treatments. Manuscript.
Lehmann, P. F. 1976. Unusual fungi on pine leaf litter induced by urea and urine. Trans. Br. mycol. Soc. 67: 251—253.
Linnemann, G. 1958. Untersuchungen zur Verbreitung und Systmatik der Mortierellen. Arch. Mikrobiol. 30: 256—267.
Lockwood, J. L. 1977. Fungistasis in soils. Biol. Rev. 52: 1—43.
Lockwood, J. L. 1988. Evolution of concepts associated with soilborne plant pathogens. A. Rev. Phytopath. 26: 93—121.
Loub, W. 1960. Die mikrobiologische Charakterisierung von Bodentypen. Die Bodenkultur A 11(1): 38—70.

Maloy, O. C. 1974. Benomyl-malt agar for the purification of cultures of wood decay fungi. Pl. Dis. Reptr 58: 902—904.
Mankau, R. 1980. Biocontrol: fungi as nematode control agents. J. Nematol. 12: 213—232.
Massari, G. 1988. Pour une méthode d'étude de la mycoflore du sol. Rev. Écol. Biol. Sol 20: 445—460.
Mishustin, E. N. (ed.). 1972. Microflora of soils in the Northern and Central USSR. Israel Progr. scient. Transl. Jerusalem.
Mishustin, E. N., O. I. Pushkinskaya, and Z. F. Teplyakova. 1961. (The ecologo-geographical distribution of microscopic soil fungi). Trudy Inst. Pochv. Alma-Ata 12: 3—64.
Morris, C. E., and D. I. Rouse. 1986. Microbiological and sampling considerations for quantification of epiphytic microbial community structure. p. 3—13. *In*: N. J. Fokkema and J. van den Heuvel (eds.) Microbiology of the phyllosphere. Cambridge Univ. Press.
Nagel-de Boois, H. M., and E. Jansen. 1971. The growth of fungal mycelium in forest soil layers. Rev. Ecol. Biol. Sol 8: 509—520.
Newell, S. Y., T. L. Arsuffi, and R. D. Fallon. 1988. Fundamental procedures for determining ergosterol content of decaying plant material by liquid chromatography. Appl. Environ. Microbiol. 54: 1876—1879.
Nordgren, A., E. Bååth, and B. E. Söderström. 1985. Soil microfungi in an area polluted by heavy metals. Can. J. Bot. 63: 448—455.
Papavizas, G. C., and R. D. Lumsden. 1982. Improved medium for isolation of *Trichoderma* spp. from soil. Plant Dis. 66: 1019—1020.
Park, D. 1968. The ecology of terrestrial fungi p. 5—39. *In*: G. C. Ainsworth A. S. Sussmann (eds.) The fungi, vol. 3. Academic Press, New York.
Park, D. 1973. A modified medium for isolation and enumeration of cellulose-decomposing fungi. Trans. Br. mycol. Soc. 60: 148—151.
Parkinson, D. 1970. Methods for the quantitative study of heterotrophic soil micro-organisms. p. 101—105. *In*: J. Phillipson (ed.) Methods of study in soil ecology. Unesco, Genève.
Parkinson, D., T. R. G. Gray, and S. T. Williams. 1971. Methods for studying the ecology of soil micro-organisms. IBP Handbook No. 19. Blackwell, Oxford.
Parkinson, D., and S. T. Williams. 1961. A method for isolating fungi from soil microhabitats. Pl. Soil 13: 347—355.
Persson, T., E. Bååth, M. Clarholm, H. Lundkvist, B. Söderström, and B. Sohlenius. 1980. Trophic structure, biomass dynamics and carbon metabolism of soil organisms in a Scots pine forest. Ecol. Bull. Stockholm 32: 419—459.
Petrini, L., and O. Petrini. 1986. Xylariaceous fungi as endophytes. Sydowia 38: 216—234 ("1985").
Petrini, O. 1986. Taxonomy of endophytic fungi of aerial plant tissues. p. 175—187. *In*: N. J. Fokkema and J. van den Heuvel (eds.) Microbiology of the phyllosphere. Cambridge Univ. Press.
Peyronel, B. 1955. Proposta di un nuovo metodo di rappresentazione grafica della composizione dei consorzi vegetali. Nuovo G. bot. ital. N. S. 62: 379—382.
Pirozynski, K. A. 1968. Geographical distribution of fungi. p. 487—504. *In*: G. C. Ainsworth and A. S. Sussman (eds.) The fungi, an advanced treatise, vol. 3. Acad. Press, London.
Plaats-Niterink, A. J. van der. 1981. Monograph of the genus *Pythium*. Stud. Mycol. 21: 242 pp.
Ponchet, J., and R. Tramier. 1971. Effets du benomyl sur la croissance de l'oeillet et la microflore des sols traités. Annls Phytopath. 3: 401—406.
Pugh, G. J. F., 1980. Strategies in fungal ecology. Trans. Br. mycol. Soc. 75: 1—14.
Pugh, G. J. F. and L. Boddy. 1988. A view of disturbance and life strategies in fungi. Proc. R. Soc. Edinb. 94B: 3—11.
Pugh, G. J. F., and J. H. van Emden. 1969. Cellulose-decomposing fungi in polder soils and their possible influence on pathogenic fungi. Neth. J. Pl. Path. 75: 287—295.

Rayner, A. D. M., and L. Boddy. 1988. fungal decomposition of wood, its biology and ecology. — J. Wiley, Chichester.
Riesen, Th., and Th. Sieber. 1985. Endophytic fungi in winter wheat. Diss. Mikrobiol. Inst. ETH, Zürich.
Rosswall, T., J. Schnürer, and S. Söderlund. 1986. Interactions of acidity, aluminium ions and microorganisms. p. 395—410. *In*: V. Jensen, A. Kjøller and L. H. Sørensen (eds.) Microbial communities in soil. Elsevier, London.
Russell, P. 1956. A selective medium for the isolation of basidiomycetes. Nature, Lond. 177: 1038—1039.
Salt, G. A. 1979. The increasing interest in 'minor pathogens'. p. 289—312. *In*: B. Schippers and W. Gams (eds.) Soil-borne plant pathogens. Acad. Press, London.
Sappa, F. 1955. La micoflora del terreno quasi elemento strutturale delle communità vegetali: I. Saggi metodologici sul Calluneto di S. Francesco al Campo (Torino). Allionia 2: 293—345.
Schnürer, J., M. Clarholm, and T. Rosswall. 1985. Microbial biomass and activity in an agricultural soil with different organic matter contents. Soil Biol. Biochem. 17: 611—618.
Siegel, M. R. 1975. Benomyl — soil microbial interactions. Phytopathology 65: 219—220.
Söderström, B. E. 1975. Vertical distribution of microfungi in a spruce forest soil in the south of Sweden. Trans. Br. mycol. Soc. 65: 419—425.
Söderström, B. E. 1977. Vital staining of fungi in pure cultures and in soil with fluorescein diacetate. Soil Biol. Biochem. 9: 59—63.
Söderström, B. E. 1979. Some problems in assessing the fluorescein diacetate-active fungal biomass in the soil. Soil Biol. Biochem. 11: 147—148.
Söderström, B. E., and E. Bååth. 1978. Soil microfungi in three Swedish coniferous forests. Holarctic Ecol. 1: 62—72.
Söderström, B. E., E. Bååth, and B. Lundgren. 1983. Decrease in soil microbial activity and biomasses owing to nitrogen amendments. Can. J. Microbiol. 29: 1500—1506.
Söderström, B. E., and S. Erland. 1986. Isolation of fluorescein diacetate stained hyphae from soil by micromanipulation. Trans. Br. mycol. Soc. 86: 465—468.
Sørensen, T. 1948. A method of establishing groups of equal amplitude in plant sociology based on similarity of species content. Kgl. Danske Vidensk. Selsk. Biol. Skr. 5(4): 1—34.
Stalpers, J. A. 1976. Identification of wood-inhabiting Aphyllophorales in pure culture. Stud. Mycol. 16: 248 pp.
States, J. S. 1981. Useful criteria in the description of fungal communites. p. 185—199. *In*: D. T. Wicklow and C. G. Carroll (eds.) The fungal community, its organization and role in the ecosystem. M. Dekker, New York, Basel.
Summerbell, R. C. 1989. Microfungi associated with the mycorrhizal mantle and adjacent microhabitats within the rhizosphere of black spruce. Can. J. Bot. 67: 1085—1095.
Sundman, V., and S. Sievelä. 1978. A comment on the membrane filter technique for estimation of length of fungal hyphae in the soil. Soil Biol. Biochem. 10: 399—401.
Swift, M. J. 1984. Microbial diversity and decomposer niches. p. 8—16. *In*: M. J. Klug and C. A. Reddy (eds.) Current Perspectives in Microbial Ecology. Am. Soc. Microbiol., Washington DC.
Swift, M. J. 1987. Organization of assemblages of decomposer fungi in space and time. p. 229—253. *In*: J. H. R. Gee and P. S. Giller (eds.) Organization of communities, past and present. Blackwell, Oxford and London.
Swift, M. J., and O. W. Heal. 1986. Theoretical considerations of microbial succession and growth strategies: intellectual exercise or practical necessity? p. 115—131. *In*: Jensen *et al.* (eds.) Microbial communities in soil. Elsevier, London.
Swift, M. J., O. W. Heal, and J. M. Anderson. 1979. Decomposition in terrestrial ecosystems. Blackwell, Oxford.

Taylor, B., and D. Parkinson. 1988. A new microcosm approach to litter decomposition studies. Can. J. Bot. 66: 1933—1939.

Thomas, A., D. P. Nicholas, and D. Parkinson 1965. Modifications of the agar film technique for assaying lengths of mycelium in soil. Nature, Lond. 205: 105.

Thornton, R. H. 1956. Fungi occurring in mixed oakwood and heath soil profiles. Trans. Br. mycol. Soc. 39: 485—494.

Trautnann, C., M. Peters, and G. Kraepelin. 1992. Die Mucorales-Flora in Streu- und Bodenhorizonten eines Berliner Kiefernwaldes. I. Einfluß einer Kalkdüngung auf die Mucorales-Population. Z. Mykol. 58: 3—14.

Tresner, H. D., M. P. Backus, and J. T. Curtis. 1954. Soil microfungi in relation to the hardwood forest continuum in southern Wisconsin. Mycologia 46: 314—333.

Tribe, H. T. 1957. Ecology of micro-organisms in soils as observed during their development upon buried cellulose film. p. 287—298. *In*: R. E. O. Williams and C. C. Spicer (eds.) Microbial Ecology. 7th Sympos. Soc. gen. Microbiol.

Tribe, H. T. 1977. Pathology of cyst nematodes. Biol. Revs 52: 477—507.

Tribe, H. T. 1980. Prospects for the biological control of plant-parasitic nematodes. Parasitology 81: 619—639.

Tsao, P. H. 1970. Selective media for isolation of pathogenic fungi. A. Rev. Phytopath. 8: 157—186.

Tsao, P. H., and S. O. Guy. 1977. Inhibition of *Mortierella* and *Pythium* in a *Phytophthora*-isolation medium containing Hymexazol. Phytopathology 67: 796—801.

Verhoef, H. A., J. E. Prast, and R. A. Verweij. 1988. Relative importance of fungi and algae in the diet and nitrogen nutrition of *Orchesella cincta* (L.) and *Tomocerus minor* (Lubbock) (Collembla). Functional Ecol. 2: 195—201.

Visser, S., and D. Parkinson. 1975. Fungal succession on aspen — poplar leaf litter. Can. J. Bot. 53: 1640—1651.

Waid, J. S., and M. J. Woodman. 1967. A method of estimating hyphal activity in soil. Pedobiologia 7: 155—158.

Wainwright, M. 1988. Metabolic diversity of fungi in relation to growth and mineral cycling in soil — a review. Trans. Br. mycol. Soc. 90: 159—170.

Warcup, J. J. 1950. The soil-plate method. Nature, Lond. 170: 166—167.

Warcup, J. H. 1951a. The ecology of soil fungi. Trans. Br. mycol. Soc. 34: 376—399.

Warcup, J. H. 1951b. Effect of partial sterilization by steam or formalin on the fungus flora of an old forest nursery soil. Trans. Br. mycol. Soc. 34: 520—532.

Warcup, J. H. 1955a. Isolation of fungi from hyphae present in soil. Nature, Lond. 175: 953.

Warcup, J. H. 1955b. On the origin of fungi developing on soil dilution plates. Trans. Br. mycol. Soc. 38: 298—301.

Warcup, J. H. 1967. Fungi in soil. p. 51—110. *In*: N. A. Burges and F. Raw (eds.) Soil biology. Academic Press, London.

Webster, J. 1970. Coprophilous fungi. Trans. Br. mycol. Soc. 54: 161—180.

Weller, D. M. 1988. Biological control of soilborne plant pathogens in the rhizosphere with bacteria. A. Rev. Phytopath. 26: 379—407.

West, A. W. 1988. Specimen preparation, stain type, and extraction and observation procedures as factors in the estimation of soil mycelial lengths and volumes by light microscopy. Biol. Fertil. Soils 7: 88—94.

Wicklow, D. T. 1981a. The coprophilous fungal community: a mycological system for examining ecological ideas. p. 47—76. *In*: D. T. Wicklow and G. C. Carroll (eds.) The fungal community, its organization and role in the ecosystem. M. Dekker, New York, Basel.

Wicklow, D. T. 1981b. Biogeography and conidial fungi. p. 417—447. *In*: G. T. Cole and B. Kendrick (eds.) Biology of conidial fungi, Vol. 1. Acad. Press, New York.

Wicklow, D. T., and G. C. Carroll (eds.). 1981. The fungal community, its organization and role in the ecosystem. M. Dekker, New York, Basel.

Widden, P. 1981. Patterns of phenology among fungal populations. p. 387—401. *In*: D. T. Wicklow and G. C. Carroll (eds.) The fungal community, its organization and role in the ecosystem. M. Dekker, New York, Basel.

Widden, P. 1986a. Microfungal community structure from forest soils in southern Quebec, using discriminant function and factor analysis. Can. J. Bot. 64: 1402—1412.

Widden, P. 1986b. Seasonality of forest soil microfungi in southern Quebec. Can. J. Bot. 64: 1413—1423.

Widden, P. 1986c. Functional relationships between Quebec forest soil microfungi and their environment. Can. J. Bot. 64: 1424—1432.

Widden, P., and D. Parkinson. 1973. Fungi from Canadian coniferous forest soils. Can. J. Bot. 51: 2275—2290.

Widden, P., and D. Parkinson. 1975. The effects of a forest fire on soil microfungi. Soil Biol. Biochem. 7: 125—138.

Wilcoxon, F., and R. A. Wilcox. 1964. Some rapid approximate statistical procedures. Lederle Lab., Pearl River, New York.

8. Communities of parasitic microfungi

GERALD HIRSCH and UWE BRAUN

Abstract

The main aspects concerning the role of parasitic micromycetes in phytocoenology as well as mycocoenology are discussed. Phytoparasitic fungi are important elements of definite phytocoenoses. Investigations on the composition of this fungal group in particular plant communities are of general interest on account of ecological and phytopathological reasons. These studies are classified as 'phytocoenological mycofloristics'. Papers dealing with such problems are summarized.

True consociations (unions) of parasitic micromycetes are known. The knowledge about them is summarized under the catchword "mycocoenology". The following aspects are included: mixed infections, phylloplane fungi, rhizosphere fungi, interactions, endophytes, hyperparasites. The special problems connected with the description of mycosynusia of parasitic fungi are discussed, and the new associations Peronosporo parasiticae — Albuginetum and Puccinio dioicae — Schizonelletum are proposed. They shall be considered as basis for further discussions on this problem.

8.1. Introduction

The biology and ecology of meta- and holobiotrophic fungi (according to Luttrell 1974) is dominantly influenced by the very close, essential connection between fungus and host plant. The environmental impact on such organisms happens at two levels. Mäkinen (1966) called these levels 'microecology' and 'macroecology'. The host represents the immediate habitat for a parasite, and the environment influences the biotrophic organisms via hosts (= microecology, Mäkinen l.c.). On the other hand, there is a direct influence

W. Winterhoff (ed.), Fungi in Vegetation Science, 225—250.

of the environmental factors (temperature, moisture, etc.) on the parasites (= macroecology, Mäkinen l.c.). Within these processes, the microclimate, that strongly depends on the phytocoenological structure of the biotops, plays an important role. Hence, Urban (1980) stressed that the ecology of parasitic fungi also includes the interrelations between the parasites and the vegetation types, which are either natural or artificial plant communities (phytocoenoses). Undoubtedly, there is a close co-evolution between hosts and their parasites, but also between biotrophic fungi and the phytocoenoses which represent their habitats (Savile 1971, Harlan 1976).

Urban (1980) and Savile and Urban (1982) discussed the importance of these processes in the evolution of the rust fungi, especially with regard to heteroecious taxa. Urban (1980) stated that the natural vegetation as well as artificial units (e.g. agrophytocoenoses) may spread as definite phytocoenoses together with their parasites. These processes occurred in the past and may still occur today. Therefore, the knowledge of the evolution of the phytocoenoses should be introduced into ecological investigations on phytoparasitic fungi.

The experiences show that the occurrence as well as the abundance of phytoparasitic fungi in a landscape do not only depend on the mere existence of the host. The various plant communities in which the hosts grow also play an important role. Braun (1978a), for instance, studied the distribution, frequency and abundance of *Puccinia aegopodii* Röhl. in various phytocoenoses of central G.D.R. This fungus occurred very frequently in nitrophilous border communities (Urtico-Aegopodietum and Alliario-Chaerophylletum temuli). 45.8% of the investigated places (together more than 100) showed infections of the host by *P. aegopodii.* But only 13.8% of mere *Aegopodium* stands and mixed plant communities composed of meadow elements as well as elements of nitrophilous border communities were attacked by this fungus. The frequency of *P. aegopodii* infections in *Aegopodium* stands in forests (e.g. Fraxino-Ulmetum, Sambuco-Ulmetum) was intermediate.

On account of the expounded facts as well as some other reasons discussed later, it is necessary and very useful to investigate the composition of the mycoflora of parasitic fungi in definite phytocoenoses. However, the mycelia of the studied parasites are not associated. The particular species usually occur on different host plants within a single, studied community. Hence, such studies cannot be classified as mycocoenology. According to Kreisel (1957, 1985), mycocoenology deals with mycocoenoses (= mycosocieties in the sense of this book) which are characterized by the consociation of different mycelia on or within a substrate. Dörfelt (1974a,b) created the appropriate term 'phytocoenological mycofloristics' to characterize investigation on the mycoflora of particular plant communities, including saprophytic fungi and mycorrhiza symbionts. He stressed that such studies represent first elements of the mycocoenology. We agree with Dörfelt (l.c.) and separate this chapter into two subchapters — 'Phytocoenological mycofloristics of phytoparasitic fungi' and 'Mycocoenology of parasitic micromycetes'.

However, it must be stated that the term 'mycocenosis' is often used in different ways. Apinis (1972), for instance, applied it to the total of fungi in a particular biocenosis, and Barkman (1976) rejected this term entirely. He clearly expressed that, in his opinion, there are no mycocenoses at all. According to Barkman (l.c.), mycocenoses are mere 'taxocenoses'.

Kreisel (1985) stated categorically that obligately phytoparasitic fungi are not a subject of mycocoenology. However, this statement is too absolute. There are, indeed, many cases of various parasitic fungi associated on or in a single host or host organ. The biological meaning as well as the mycocoenological aspects of such associations will be discussed in 8.3.

8.2. Phytocoenological mycofloristics of phytoparasitic fungi

8.2.1. *General considerations*

Analyses of the mycoflora of phytoparasitic fungi in definite plant communities provide a wide range of scientific information on various aspects of mycology, phytopathology and phytocoenology, e.g.

- Data concerning the evolution and taxonomy of parasitic fungi; co-evolution of host plants and host plant communities and their parasites (e.g. Savile 1971, Urban 1980, Savile and Urban 1982, Scholz 1976).
- Data on the degree of anthropogenic influence on landscapes and their biocoenoses (Scholz 1976).

The occurrence and the frequency of parasitic fungi in phytocoenoses reflect the impact of man. They may be used as indicators of these influences. In natural phytocoenoses, there are usually balances between the plant elements and their parasites. Both parts passed through a co-evolution. In anthropogenic communities, this balance is often disturbed. Dinoor and Eshed (1984) generally stated that disease levels reflect the degree of stability in ecosystems. Low levels indicate stability. Epidemics are often the result of anthropogenic impacts or other changes of the external conditions (Burdon and Shattock 1980). Severe epidemics in natural phytocoenoses may cause ecological disasters as well (e.g. Dutch elm disease, eucalyptus dieback in Australia). Dinoor and Eshed (l.c.) even stated that some epidemics in natural communities may be a part of the natural phytocoenosis balance and that the result will be much worse if disturbance occurred.

- Data on the distribution of parasitic fungi in the landscape; mycofloristical characterizations of particular phytocoenoses; interrelations between the parasites of various plant communities in definite landscapes; interrelations between agrophytocoenoses (including the cultivated hosts and the surrounding phytocoenoses (e.g. Braun 1982).
- Data on the effects of parasites on the structure of phytocoenoses (Harlan 1976, Burdon and Shattock 1980, Burdon 1982).

Parasitic fungi may obviously influence the structure and development of plant communities; e.g., they are able to eliminate plants in thick uniform stands (Harlan 1976) or they can cause premature leaf fall. Such effects may lead to a clearing of the phytocoenosis and, hence, to successions. Furthermore, pathogens influence the competition of the plant elements of a community and the distribution in the ecosystems (Chilvers and Brittain 1972).

Ellenberg (1978) referred to the fact that various 'snow mould fungi', e.g. *Herpotrichia nigra* Hartig, may influence the competition between conifers and deciduous trees in European subalpine forests (e.g. sycamore-beech forests) under oceanic condition, in mild winters.

The action of phytoparasitic fungi may also yield substrates and habitats for other fungi and fungal successions may be introduced. Kreisel and Müller (1987) described the Pleurotetum cornucopiae, a fungus association at dead trunks of elms — a consequence of the elm disease.

Dinoor and Eshed (1984) provided a survey on the importance of parasitic fungi in plant communities based on phytopathological viewpoints. They noted the following grounds for the growing interest in diseases of natural phytocoenoses:

- they are potential sources of inoculum, of new biotypes and of new diseases;
- they are important in the search for disease resistance (e.g. Leppik 1970);
- their knowledge is essential in the search for defense strategies of wild plants and for biocontrol agents.

8.2.2. *Survey of the phytocoenological-mycofloristical investigations*

There are numerous papers dealing with phytoparasitic and other fungi of particular landscapes, national parks, islands, etc. Some of these works provide descriptions of the vegetation of the landscapes concerned or they refer to special papers dealing with the vegetation of the studied areas, e.g. Durrieu (1966), Sandu-Ville *et al.* (1973), Dennis *et al.* (1977), Cummins (1979), Geljuta (1979a,b), Tanda (1983), Parmelee (1984) and Parmelee and Ginns (1986). Unfortunately, the mentioned papers as well as numerous similar works do not assign the listed particular collections to any plant community. The number of papers exactly dealing with the pathogenic fungi of definite phytocoenoses is rather limited. There are especially some authors in Poland that have been engaged in such studies. These authors used different methods. Majewski (1967) investigated the mycoflora of the 'Kampinos Forests' (Poland) and analysed the pathogenic fungi which attacked the character species of the phytocoenoses of the areas concerned. However, the complete list of collected species does not contain any information on the plant communities where the material had been gathered.

Majewski (1971) published a detailed work on the phytoparasitic fungi of the 'Bialowieza National Park' (Peronosporales, Erysiphales, Uredinales, Ustilaginales). He described and listed the fungi of various phytocoenoses. These studies represent mycofloristical characterizations of numerous plant communities. Furthermore, he analysed and compared the quantitative and qualitative peculiarities of the neighbouring associations. On account of his investigations on forests, he stated that the richest mycoflora of parasitic fungi had been found in the Circaeo-Alnetum as well as in the Querco-Carpinetum, whereas the number of these fungi was fairly small in coniferous forests. According to Majewski (l.c.), it seems that these differences are owing to different moisture conditions in the mentioned plant communities. Numerous potential host plants in the Querco-Piceetum and Pino-Quercetum, also growing in the Circaeo-Alnetum and Querco-Carpinetum, are only infected in the latter associations. The occurrence of the Peronosporales was largely limited to moist places in deciduous forests. The powdery mildews most richly occurred in the Circaeo-Alnetum. Rusts and smuts were most abundant in the Querco-Carpinetum. It is indicated by these results that the ecological conditions in deciduous forests (with regard to the studied fungal groups) favour the development of parasitic fungi (compared with coniferous forests). The degree of the similarity of the mycoflora of parasitic fungi in various phytocoenoses depends on the floristic similarity of the plant communities.

Kućmierz (1973, 1977) dealt with plant parasitic fungi of the Ojców

National Park and the Pieniny Mts. (Poland), respectively. This author found the greatest diversities of parasitic fungi in meadow and pasture plant communities, synanthropic associations as well as riverside phytocoenoses. He pointed out obvious differences in the abundance of phytoparasitic fungi in various forest communities. Kućmierz (1973) stated that there are significant differences — with regard to the number of collected fungus species — between 'dry' and 'moist' phytocoenoses. The moister the plant community the richer the flora of parasitic fungi. In 1977, this author carried out qualitative as well as quantitative investigations on vascular plants and their parasites in numerous associations. He estimated the abundance of the fungi in the particular phytocoenoses and the degree of the damage of each host species. Kućmierz (1977) was the first author to introduce quantitative aspects on the occurrence of phytoparasitic fungi in plant communities.

Romaszewska-Sałata (1976, 1977) investigated xerothermic plant communities near Lublin (Poland). She used the same methods as Majewski (l.c.). Most phytoparasitic fungi were found in the following associations: Thalictro-Salvietum pratensis, Inuletum ensifoliae and a *Brachypodium pinnatum-Teucrium chamaedrys* association. Similar studies were carried out by Romaszewska-Sałata (1981). In this paper, she dealt with the xerothermic phytocoenoses of the Malopolska Upland. The collected parasitic fungi were summarized in a list and assigned to the particular plant communities where they had been found.

Mułenko (1981) studied the microscopic phytoparasitic fungi in the forest — peat-bog — reservation at Brzeziczno (Poland). He analysed 14 associations, but only in 11 phytocoenoses parasites could be traced. The names of the species and their host plants are listed in a table and the occurrence in the forest or peat-bog communities is indicated. Furthermore, he added the data of collecting (month) of the particular taxa. Mułenko's (l.c.) investigations include quantitative consideration on the occurrence of the fungi concerned. He used the following symbols: + = sporadically; ++ = frequently; and +++ = very frequently appearing. The number of species and their frequency and abundance were fairly rich in the forest communities but poor in the peat-bogs.

The work of Braun (1982) includes the results of extensive investigations on phytoparasitic fungi in numerous phytocoenoses (forests, meadows, agrophytocoenoses, ruderal communities, etc.) carried out in various landscapes of central and southern G.D.R., ranging from the plains to the mountains.

8.2.3. *A qualitative-quantitative method for phytocoenological-mycofloristical investigations*

Braun (l.c.) tried to introduce new methods, especially with regard to the quantitative aspects of the fungal occurrence. This author checked the entire areas of the particular plant communities in order to get complete lists of all occurring parasites. With regard to the frequency of the parasites, it must be taken into consideration that there is a difference between the absolute and the relative frequency. The absolute frequency considers the number of parasitized plants in the community and it depends on the abundance of the host. A parasitic fungus may be absolutely frequent in a phytocoenosis provided that the host is fairly abundant. The absolute frequency of a parasite on a host which is rather rare in a definite plant community cannot be large, even when nearly all individuals are attacked. In the latter case, the relative frequency (referred to the extant number of plants in the community, independent of their abundance) would be high. On the other hand, it is possible that a parasitic fungus in a special phytocoenosis possesses a high absolute frequency, but only a moderate relative frequency (host very abundant in the community, parasite fairly frequent, but only a part of the very numerous potential host plants infected). Hence, the differentiation between absolute and relative frequency was introduced into the quantitative considerations.

The following categories for the frequency of the fungal occurrence were proposed (based on estimated values):

++ = Severe, dominant occurrence in the phytocoenosis, relative as well as absolute frequency large, often more than 30% of the hosts infected.

\+ = Occurrence moderate, often scattered, incomplete, always less than 30% of the hosts infected, relative frequency also moderate.

+(+) = The same occurrence as in '+', but relative frequency large (host not very frequent in the studied phytocoenosis).

+− = Absolute as well as relative frequency in the phytocoenosis small, always less than 5% of the hosts infected.

+(−) = Absolute frequency small, but relative frequency large.

−(+) = Single occurrences (host abundant in the phytocoenosis).

−(−) = Single occurrences (host rare in the phytocoenosis).

The terms abundance and frequency in the phytocoenological sense cannot be used for the parasitic fungi in the same way as for vascular plants and macrofungi. 'Frequency' refers to the portion of attacked individuals of a host plant in a single investigated place. 'Abundance' would refer to the degree and the severeness of the infections on the particular host individuals. But 'abundance' is not applied in the present studies for the parasites, but only for the host plants in the common phytocoenological sense. The term 'dominant' ('dominance'), with regard to the phytoparasitic fungi, is also not used in the common phytocoenological manner. It only means that parasites with the highest frequencies are very characteristic for the plant communities concerned; they are dominant, compared with the other taxa.

In phytocoenoses which are rather limited relating to space, the entire areas were used for the estimations of the frequency. Expanded communities, e.g. woods, large meadows and agrophytocoenoses, were separated into smaller units (ca 10 × 10 m) and the estimations were carried out on these places.

In order to obtain a general information about the characteristic stocks of phytoparasitic fungi of the plant communities, it was necessary to compare and to condense the data of the numerous particular investigations carried out in the same phytocoenoses, but in different landscapes. To get a manageable value, a 'frequency figure' was introduced.

Estimation of the frequency in particular investigations		Corresponding particular frequency figure (PFF)
++	=	4
+	=	3
+(+)	=	3
+−	=	2
+(−)	=	2
−(+)	=	1
−(−)	=	1

The condensed 'mean frequency figure' of the occurrence of a parasite in a phytocoenosis (MFF) represents the average of all PFF:

$$\text{MFF} = \frac{\sum \text{PFF}}{\text{NP}}$$

(NP = number of investigated places)

The MFF may range between 4 and 0. The higher the figure the higher the frequency and dominance of the parasite. On account of the MFF, three frequency categories were established:

Category I — MFF > 1 (−4), parasitic fungi very frequent, characteristic and dominant in the plant community.

Category II — MFF 0.5—1, parasites regularly occurring in the plant community, but not frequent and not dominant.

Category III — MFF < 0.5, parasites only occasionally occurring in the plant community.

The agrophytocoenosis Euphorbio-Melandrietum, for instance, was characterized as follows (cultivated hosts excluded):
Category I — Puccinia recondita on *Agropyron repens* (MFF 3), *P. punctiformis* on *Cirsium arvense* (MFF 2.7), *Blumeria graminis* on *Agropyron repens* (MFF 2.2), *Peronospora chenopodii* on *Chenopodium album* (MFF 1.4) and *P. aparines* on *Galium aparine* (MFF 1.1).
Category II — *Melampsora euphorbiae* on *Euphorbia helioscopia* (MFF 0.6), *Puccinia poae-nemoralis* on *Poa annua* (MFF 0.5), *P. absinthii* var. minor on *Artemisia vulgaris* (MFF 0.5), *Erysiphe galii* var. *galii* on *Galium aparine* (MFF 0.9), *E. galeopsidis* on *Lamium amplexicaule* (MFF 0.7), *E. polygoni* on *Polygonum aviculare* (MFF 0.5), *Peronospora sisymbrii-of-ficinalis* on *Sisymbrium officinale* (MFF 0.6), *P. brassicae* on *Sinapis arvensis* (MFF 0.5), *P. sisymbrii-sophiae* on *Descurainia sophia* (MFF 0.5), and *Plasmopara leptosperma* on *Matricaria maritima* ssp. *inodora* (MFF 1). The category I and II species comprise 24.4% of the total of the collected parasitic fungi.

The limited space of the present chapter does not permit to discuss all investigated plant communities in detail. However, in Table I a list of papers dealing with phytocoenological mycofloristics is provided accompanied by the names of the studied phytocoenoses.

With regard to Braun's (l.c.) investigations, the following general results and remarks should be added.

- The mycofloristic characterization of the widely distributed phytocoenoses (concerning the frequent, dominant species of category I and II) was very uniform in the entire area, in different landscapes, often even from the plains to the mountains.
- Phytocoenologically related associations, e.g. Euphorbio-Melandrietum/ Aphano-Matricarietum or Dauco-Arrhenatheretum/Alchemillo-Arrhenatheretum, possess very similar stocks of parasites (correlated with the close phytocoenological affinity). The category I and II species are largely agreeing. Separations on account of the parasitic fungi are hardly possible.
- There is a conspicuous pauperization of the mycoflora in nearly all phytocoenoses with rising altitude (from the plains to the mountains).
- There are numerous interrelations and exchanges between the stocks of parasites in the various neighbouring plant communities of a landscape; e.g. about 75% of the dominant parasitic fungi (category I and II) of the agrophytocoenoses in the studied areas also occurred in the neighbouring communities (meadows, ruderal vegetation, etc.), a third even in category I and II, as well.

Table I. List of the mycofloristically-phytocoenologically investigated plant communities

Author	Plant communities
Majewski (1967)	Caricetum elatae, Caricetum paradoxae, Carici elongatae-Alnetum, Circaeo-Alnetum, Glycerietum maximae, Pino-Quercetum, Tilio-Carpinetum, Vaccinio uliginosi-Pinetum (furthermore, general considerations on Corynephoretalia, Potametea, Phragmitetea, Secalino-Violetalia arvensis, Onopordetalia).
Majewski (1971)	Carici elongatae-Alnetum, Circaeo-Alnetum, Pino-Quercetum, Peucedano-Pinetum, Querco-Carpinetum medioeuropaeum, Querco-Piceetum, Sphagnetum medii, Salici-Franguletum, Vaccinio uliginosi-Pinetum.
Kućmierz (1973)	Alno-Padion, Arrhenatheretum elatioris, Corylo-Peucedanetum cervariae, Fagetum carpaticum, Festucetum pallentis, Geranio-Petasitetum, Glycerietum plicatae, Junco-Menthetum longifoliae, Lamio-Veronicetum politae, Lolio-Cynosuretum, Origano-Brachypodietum pinnati, Phyllitido-Aceretum, Pino-Quercetum, Scirpo-Phragmitetum, Tilio-Carpinetum, Vicietum tetraspermae.
Romaszewska-Sałata (1976)	*Brachypodium pinnatum-Teucrium chamaedrys* ass., Inuletum ensifoliae, Koelerio-Festucetum sulcatae, Peucedano cervariae-Coryletum, Prunetum fruticosae, Stipetum capillatae.
Kućmierz (1977)	Alnetum incanae, Anthylli-Trifolietum montani, Arrhenatheretum elatioris, Carici-Fagetum, Cirsietum rivularis, Fagetum carpaticum, Gladiolo-Agrostetum, Laserpititum latifolii, Phyllitido-Aceretum.
Romaszewska-Sałata (1981)	*Brachypodium pinnatum* ass., Inuletum ensifoliae, Koelerio-Festucetum sulcatae, Peucedano cervariae-Coryletum, Prunetum fruticosae, Sisymbrio-Stipetum achilleetosum pannonicae, Sisymbrio-Stipetum botriochloëtosum, Sisymbrio-Stipetum poëţosum bulbosae, Seslerio-Scorzoneretum purpureae, Scabioso-Teucrietum, Thalictro-Salvietum pratensis.
Mułenko (1981)	Caricetum limosae, Caricetum strictae, Carici-Agrostidetum, Juncetum effusi, Molinietum caeruleae, Molinio-Pinetum, Pino-Quercetum, Querco-Piceetum, Scirpo-Phragmitetum, Sphagnetum medii, Salici-Franguletum, Vaccinio myrtilli-Pinetum, Vaccinio uliginosi-Pinetum.
Braun (1982)	Aegopodio-Sambucetum, Alchemillo-Arrhenatheretum, Aphano-Matricarietum, Arrhenathero-Artemisietum, Atriplicetum nitentis, Balloto-Malvetum sylvestris, *Calamagrostis epigeios* ass., Chenopodietum boni-henrici, Chenopodio-Ballotetum nigrae, Convolvulo-Agropyretum, Dauco-Arrhenatheretum, Dauco-Picridetum, Descurainio-Atriplicetum oblongifoliae, Echio-Melilotetum albi, Epilobio-Senecietum sylvatici, Euphorbio-Melandrietum, Festuco-Brachypodietum, Filipenduletum ulmariae, Galio-Carpinetum, Ligustro-Prunetum, Luzulo-Fagetum, Luzulo-Quercetum petraeae, Lycietum halimifolii, Onopordetum acanthii, Plantagini-Lolietum, Plantagini-Polygonetum avicularis, Polygono-Cirsietum, Pontentilletum anserinae, Rubietum idaei, Sambucetum nigrae, Senecietum fuchsii, Tanaceto-Artemisietum, Tussilaginetum farfarae, Urtico-Aegopodietum, Urtico-Artemisietum vulgaris, Urtico-Malvetum.

(nomenclature of the associations according to the particular authors)

- Many plant communities in agriculturally exploited landscapes house a large number of parasites that represent potential sources of infections for the cultivated plants.
- Phytoparasitic fungi are often especially rich in border plant communities (border zones to other phytocoenoses). The same observation has been made by Kućmierz (1973).

8.3. Mycocoenology of parasitic micromycetes

8.3.1. *General considerations*

The living plant represents a potentially available substrate for fungi, and is thus on principle comparable with all other substrates utilized by them. Indeed, there is a very great number of phytoparasitic fungi, and most probably they constitute the greater part of all hitherto described species. We may state that nearly all green plants are attacked by fungal parasites. Mostly only one species occurs on an infected plant part. Frequently, however, consociations of more than one parasite appear. Under the supposition that the mycelia compete for the substrate, these consociations are to be considered as mycosynusiae (= true mycocoenoses in the sense of Kreisel (1957, 1985)). There is no good reason to refuse them as subjects of mycocoenology, as Kreisel (1985: 71) has done. The same is meant for consociations of parasitic micromycetes on other substrates: the community, for instance, of more than one dermatophyte on the human skin should be called a mycocoenosis! The parasitism of fungi on other fungi (mycoparasites, hyperparasites) represents a special case. This set of problems will be dealt with later.

Hitherto consociations of phytoparasitic fungi have never been examined under a specific mycocoenological point of view, apart from the results of the phytocoenological mycofloristics dealt with in the preceding paragraph. Hence it follows that there are no specific methods for the examination and adequate description of phytoparasites' mycosynusiae. The phytocoenological methods of the Braun-Blanquet school as well as mycocoenological methods developed for macromycetes by different authors (see chapter 2 in this handbook!) are not applicable to parasitic micromycetes. It is, therefore, not surprising that there is no association (or better union, cf. 1.10.4) of phytoparasitic fungi formally described till now.

On the other hand, there is a remarkable increase of publications, especially in the phytopathological literature of the last years, which deal with the interrelationships between different fungi on plants, pathogenic or not.

Details will be stressed later. Some of these works could be called 'coenological', even if the primary intentions of the particular authors did not aim at these problems. The background of this development is the rising requirement for biological control of economically important plant parasites and vermins. The situation needs a raising knowledge about autecology and synecology of the parasites as well as their coenological relations.

8.3.2. *Mixed infections*

Consociations of obligate fungal phytoparasites are always poor in species. In general, they consist of two species, rarely more. It is only known for a few of these communities that they occur more frequently or even regularly in a special area. Other consociations of parasites are more accidental and occur extremely rare, hence they are not to be considered as associations (unions). In many cases the present data are so sparse that it is not possible to estimate whether an observed fungal community on a living plant is accidental or whether it occurs regularly. In future, more attention must be paid to these questions.

Perhaps the best known case of a consociation of two different phytoparasites is the close community of the Peronosporales species *Albugo candida* (Fr.) O. Kuntze and *Peronospora parasitica* (Pers.: Fr.) Fr. on *Capsella bursa-pastoris* L., which is well documented by numerous literature records (e.g. Greville 1824, Berkeley 1836, Magnus 1894, Buhr 1956, Gustavsson 1959, etc.). A formal description of this consociation as a new association (union) follows.

> It should be noted that it is not easy to fulfil the conditions of the 'Code of Phytosociological Nomenclature' (Barkman *et al.* 1986) for describing a new association in the case of phytoparasite communities. Article 7 claims for the species concerned that one has to provide "... a quantitative indication of the quantity of their occurrence ..." This is a problem and cannot be resolved fully at the moment. Therefore, the two examples given below have to be considered as proposals only.

Relevé (1) is the holotype of the new-described association. It can be observed in summer and autumn on all epigeous parts of the host (leaves, stems, inflorescences, fruits). The hitherto known geographical distribution of the association includes the whole of Europe. Because not only the host but also both parasites have a very wide, perhaps worldwide distribution, we may suppose that the Peronosporo parasiticae-Albuginetum also occurs in other

Table II. Peronosporo parasiticae — Albuginetum ass.nov. on *Capsella bursa-pastoris* L.

	Percentage of the total of plant surface directly covered by the parasite				
Species	(1)	(2)	(3)	(4)	(5)
Albugo candida	30	5	2	50	10
Peronospora parasitica	25	2	2	10	2

(1) 14.6.1977; Romania, Iaşi; Hirsch.
(2) 28.6.1983; G.D.R., Gera; Zündorf.
(3) 29.6.1946; Germany, Franconia, Hersbruck; Starcs.
(4) 21.6.1940; Germany, Saxonia, Limbach-Oberfrohna; Ebert.
(5) 13.6.1879; Germany, Schleswig, Kappeln; Fuchs.
The parasitized specimens of *Capsella bursa-pastoris* are deposited at JE.

continents. Buhr (1956) states that some other *Capsella* species show a similar susceptibility to both parasites. However, he did not clearly express that he found them associated on the other hosts, too. Hence it is not clear whether the described association (union) is restricted to *Capsella bursa-pastoris* or not. In the mycological literature, reports can repeatedly be found about a consociation of *Albugo candida* with *Peronospora* on other hosts than *Capsella* (compare, for instance, Magnus 1894!). Such cases were observed on the following members of Cruciferae: *Brassica nigra* (L.) Koch, *Erysimum cheiranthoides* L., *Raphanus raphanistrum* L., *Sinapis arvensis* L., *Sisymbrium altissimum* L. (nomenclature of all mentioned phanerogams in this chapter follows Rothmaler 1982 resp. Tutin *et al.* 1964—1980). The *Peronospora* parasites of these plants are usually considered as specifically distinct from *P. parasitica* (Gäumann 1923, Gustavsson 1959). Observations on such consociations are relatively sparse; hence it is to be assumed that they are accidental. Should one of them occur more constantly in a region, it will be necessary to prove whether this is a subassociation (subunion) of the ubiquitous Peronosporo parasiticae-Albuginetum or an association (union) of its own. The last case would imply considerations on syntaxa of higher rank.

We have dealt with the *Peronospora-Albugo* consociation at some length on purpose, because this is the most obvious and best known case of a parasite community in Europe. There are much more consociations, often called 'mixed infections', of phytoparasites reported in the literature, but the information is extremely scattered. It is not possible to provide a complete list of such phenomena in the present survey. Some examples from Europe are based upon the personal experiences of the authors and a few literature records.

Rust + rust – infections: fairly common, for instance:

- *Uromyces cristulatus* Tranzsch. + *U. tinctoriicola* Magn. on *Euphorbia seguierana* Necker (Hirsch 1982);
- *Uromyces sublevis* Tranzsch. + *Aecidium euphorbiae*(Gmel.) Pers. on *Euphorbia nicaeensis* All. subsp. *glareosa*(Pall. ex Bieb.)A.R.Sm. (Hirsch 1982);
- *Ochropsora ariae*(Fuck.)Ramsb. + *Tranzschelia anemones*(Pers.) Nannf. on *Anemone nemorosa* L. (Poelt 1981 and pers. obs., repeatedly);
- *Puccinia poarum* Niels. + *Coleosporium tussilaginis*(Pers.) Berk. on *Tussilago farfara* L. (pers. obs., repeatedly).

Rust + downy mildew – infections:

- *Melampsorella symphyti* (DC.)Bubák + *Peronospora symphyti* Gäumann on *Symphytum tuberosum* L. (Poelt 1981: "Mischinfektionen(. . .)sind auffällig oft zu finden."!).

Rust + smut – infections:

- *Uromyces ficariae*(Schum.)Fuck. + *Entyloma ficariae* F.v.Waldh. on *Ranunculus ficaria* L. (Poelt 1981);
- *Puccinia dioicae* Magn. + *Schizonella melanogramma*(DC.) J. Schröter on *Carex ornithopoda* Willd. (Hirsch 1985; see below!);
- *Puccinia arenariae*(Schum.)Wint. + *Microbotryum* (*Ustilago*) *violaceum* (Pers.: Pers.)G. Deml & Oberw. on *Silene alba*(Mill.) E.H.L. Krause (pers. obs., once).

Rust + rust + smut – infections:

- *Uromyces ficariae* + *Uromyces poae* Rabenh. + *Entyloma ficariae* on *Ranunculus ficaria* (Poelt 1981).

Rust + endomycete + downy mildew – infection:

- *Puccinia aegopodii* Röhl. + *Protomyces macrosporus* Ung. + *Plasmopara aegopodii* (Casp.)Trott. on *Aegopodium podagraria* L. (pers. obs., once).

Rust + powdery mildew – infections:

- many examples and references given by Blumer (1967);
- *Puccinia punctiformis*(Str.)Röhl. + *Erysiphe montagnei* Lév. on *Cirsium arvense*(L.)Scop. (Dörfelt and Braun 1977).

Smut + smut – infection:

- *Anthracoidea subinclusa*(Körn.)Bref. + *Farysia thuemenii* (F.v.Waldh.) Nannf. on *Carex riparia* Curt. (pers. obs., once).

Smut + downy mildew – infection:

- *Entyloma ficariae* + *Peronospora ficariae* Tul. on *Ranunculus ficaria* (Poelt 1981).

One of the mentioned examples, the consociation of *Puccinia dioicae* and *Schizonella melanogramma* on *Carex ornithopoda*, was frequently observed in Thuringia. It is described here as an association (union) of its own. The quantitative indication of the occurrence of the parasites on the host is given by the percentage of the total of leaf and stem surface, defined by a supposed line around the surface area, which is covered by the sori and telia (and/or uredinia) of the parasites, respectively. Of course, this can only be an estimation. It is not useful in this case to ascertain the percentage of the total of plant surface which is directly covered by the fungal spores.

Relevé (4) is the holotype of the new association (union). *Carex ornithopoda* is regularly attacked by the parasites, either by one alone or by both together. When a plant is infected by both, then *P. dioicae* primarily settles the outer leaves of the rosette, whereas *Sch. melanogramma* is to be found on the inner ones and on the flowering stems. Only rarely the spore heaps of both fungi occur in a close neighbourhood. This fact indicates specific mycelial interactions in the host as well as balanced concurrence relations, also because the infections of both are systemic. However, the infection by these parasites does not seem to be very harmful for *C. ornithopoda*, which is not obviously handicapped in its flowering and fruiting. Field observations provide evidence for the suggestion that both fungi are overwintering in their host.

Braun (1978b) has stated that the aecial host of the *Carex ornithopoda* — rust is not known yet. Since that time, a careful search for the host partner failed to be successful. Hence, we consider *Puccinia dioicae* on *Carex ornithopoda* as a race of that fungus with a *reduced* life cycle (hemi-form). As stated above, the fungus seems to overwinter in its host, and new infections could well happen by uredospores.

Hitherto the Puccinio dioicae — Schizonelletum is not known outside of Thuringia. *Carex digitata* L., a species related to *C. ornithopoda*, can also be attacked by *Schizonella melanogramma* and *Puccinia dioicae.* A consociation of these fungi on this host is, however, unknown.

Some of the specimens of *C. ornithopoda*, which bear the holotype of the new association, were furthermore attacked by the smut *Anthracoidea irregularis*(Liro ex)U. Braun & G. Hirsch. Notwithstanding, this fungus does not belong to the Puccinio dioicae — Schizonelletum. According to Vánky (1985), mycelium and sorus formation in the genus *Anthracoidea* is strongly restricted to a single ovary. Although *Sch. melanogramma* can extend its infection into the glumes, it never attacks the ovary. Therefore, an actual consociation between *A. irregularis* and *Sch. melanogramma* does not take place.

The interrelations between rust fungi and some species of the hyphomycete genus *Ramularia*, e.g. *R. coleosporii* Sacc., *R. uredinis*(Voss)Sacc. and

Table III. Puccinio dioicae – Schizonelletum ass.nov. on *Carex ornithopoda* Willd.

	Percentage of the total of plant surface that is defined by a supposed line around the spore colonies of the parasite				
Species	(1)	(2)	(3)	(4)	(5)
Schizonella melanogramma	70	50	60	40	30
Puccinia dioicae	5	2	5	35	20

(1) 1.5.1977; G.D.R., Jena, Rautal; Hirsch.
(2) 9.5.1979; G.D.R., Jena, Sonnenberge; Hirsch.
(3) 27.5.1979; G.D.R., Weimar, Buchfart; Manitz.
(4) 20.5.1980; G.D.R., Jena, Hummelsberg; Hirsch.
(5) 9.5.1981; G.D.R., Jena, Mühltal; Manitz.
The parasitized specimens of *Carex ornithopoda* are deposited at JE.

R. uredinearum Hulea, are still unclear. Most authors consider these species as hyperparasites on rusts, e.g. *Coleosporium, Melampsora, Aecidium.* Petrak (1927) doubted that these taxa represent true rust parasites. He stated that *R. coleosporii* races on various *Coleosporium* species fully agree with the corresponding *Ramularia* taxa on leaves of the same green host plants. Furthermore, *Ramularia uredinis* on *Melampsora* coincides morphologically entirely with *Ramularia rosea*(Fuck.)Sacc. on the same *Salix* hosts. Petrak (l.c.) considered '*R. uredinis*' as a consociation of *R. rosea* with *Melampsora*. There is a similar situation in '*R. uredinearum Hulea*' which must be regarded as a consociation of *Ramularia cerinthes* Hollós and *Aecidium asperifolii* Pers.

8.3.3. *Phylloplane fungi*

Leaf surfaces of green terrestrial plants usually possess a richly developed mycoflora, which often consists of a dozen of different micromycetes or more. These phylloplane fungi are especially well and conspicuously developed in the tropics. Most of them are considered as saprobes, although at least two voluminous groups consist partly (sooty moulds) or completely (powdery mildews) of obligate phytoparasites. The same can also be stated for many deuteromycetes belonging to this group. However, it must vigorously be pointed out that just in the phylloplane fungi a division into parasites and saprophytes is often neither possible nor useful. Many apparent saprophytes are actually facultative parasites, whose hyphae can penetrate the host under determined conditions. Besides a special disposition of the host, such conditions can also be constituted by atmospheric influences or the primary infection of a simultaneously present obligate parasite. In addition to this, Sörgel (1957) revealed that seemingly many epiphyllous fungi are obviously present also inside the apparently healthy leaf in their mycelial form.

The stressed facts indicate why the communities of phylloplane fungi are shortly treated in this chapter, even although they are not generally parasitic in a strict sense. The greater part of our knowledge about these fungi was summarized in the book of Blakeman (1981), particularly from an ecological point of view. Blumer (1933 and 1967) apparently is the first author to specify consociations of phylloplane fungi from a coenological viewpoint, whilst considering accompanying fungi of powdery mildew colonies on leaf surfaces. He writes "... Lebensgemeinschaft auf kleinstem Raum ..." (1933: 39) and even gives the title 'Die Mehltau-Biozönose' for a chapter of its own in his book of 1967 (106 ff.). He lists some genera of mostly imperfect fungi, whose representatives were more or less regularly found to be associated with powdery mildews. Blumer even speaks of a 'Sukzession' which leads finally to a total destruction of the epiphyllous mildew mycelium by bacteria.

Certainly there are true, perhaps host-specific associations of epiphyllous fungi, and they are waiting for being correctly described. Members of these fungal communities are not only true parasites but also saprophytes, at which some of them can secondarily change to a parasitic behaviour, either on the host plant or on other fungi (see below, hyperparasites!). The interrelationships between organisms on the leaf surface seem to be very complex. They are still more complicated when we take into consideration additional groups of organisms (e.g. animals, bacteria), which influence the fungal population in its structure and dynamics.

An important recent work on the phylloplane fungal communities of *Lolium perenne* L. is that of Thomas and Shattock (1986). The authors provide a detailed examination of filamentous fungal consociations (the yeasts represent a further important group of epiphyllous fungi), based on extensive mathematical analyses of the results won by direct observation of sampled and incubated leaves. They were able to distinguish seven groups of recurrent species made up of three pathogen groups, each with an associated saprophyte, and four groups of primary and secondary saprophytes. Although the groups were not described as associations, they may well represent some. This study could serve as an example for similar works.

Only a little can be said about phylloplane fungal communities in the tropics.

At least under humid conditions the leaves of many plants usually are covered with a thick and conspicuous mat of mostly dark pigmented fungal mycelia. In this respect the so-called ‘sooty moulds’ (members of the ascomycetous order Dothideales) play a dominant role. Hansford (1946) studied and described a lot of these fungi, mainly from tropical Africa. He divided them into three groups: 1. true parasites on higher plants; 2. pure saprophytes, which live on insect excrements etc.; and 3. hyperparasites on fungi of either group 1. or 2. Although the author gives many single notes on consociations of different fungi, no definite coenological conclusions can be drawn from his work. It is unknown whether true mycosynusiae with stable species combinations and host relations do occur amongst these fungi.

One possibly important fact should be mentioned. Sörgel (1957) and recently Sutton (1986) referred to the phenomenon that in the case of more than one fungus on a leaf, the species are not equally distributed over the leaf surface. They rather form colonies, which usually do not show any mutual penetration but inhibitory effects. Such effects can even result in the formation of black lines of disappeared hyphae and excretions which separate the colonies (very convincingly illustrated by Sutton l.c. with leaves of *Banksia integrifolia*). This line formation is well known from dual cultures in the laboratory and is comparable with the black crust formation in stumps caused by white-rot fungi during mutual inhibition. The coenological meaning of this is not quite clear: can phylloplane fungi, which inhabit different areas on a leaf and show inhibitory effects to each other, belong to one and the same mycosociety? This question has been answered in the above mentioned comparable case de facto with ‘Yes’ by the mycocoenology of wood-inhabiting macromycetes, and we hold the same opinion for the phylloplane fungi.

Similar aspects as to the phylloplane should be applied to the stem surface and the surface of other epigeous parts of the green plant. However, the present knowledge about the mycoflora of these plant parts is very poor. Partly, it will be similar to that of the phylloplane, but stems, flowers, fruits and especially the bark of woody plants certainly have their own fungi and their own mycosocieties. As an example, we would like to refer to the study of Webber and Hedger (1986) on elm bark fungi. Although partly experimental, this work gives some insights into the structure and dynamics of an elm bark fungal community of mostly saprobes, in which the well-known parasite *Ceratocystis ulmi* (Buism.) C. Moreau is overwintering and living as a saprophyte, too, but during the winter months only.

8.3.4. *Rhizosphere fungi*

There is a vast amount of literature dealing with the fungi associated with plant roots. Besides more randomly present soil saprophytes, mycorrhizal fungi can regularly be found as symbiontic partners of the plant at the roots, and more or less aggressive parasites, too. It is not to be supposed that there are specialized parasite communities. There is rather a complex of species with different nutritional requirements, the composition of which is depending on external factors (especially the type and the structure of soil) and the host species. Undoubtedly the structure of this species complex exhibits important fluctuations and shows great dynamics in dependence on temperature, humidity, fertilization of soil, health condition of the host, etc. Furthermore, seasonal changes are common. Therefore, we do not see any practicable base for a successful application of coenological methods on fungi of this sphere. After all, it seems to be impossible to define relatively stable mycosynusiae in the rhizosphere.

8.3.5. *Interactions*

Mechanisms of parasite-plant-interactions are the subject of phytopathological investigations for a long time. Research has revealed a lot of details, even on the molecular level (e.g. the phytoalexine system of higher plants inhibitory to the invading parasite). However, it is largely unknown what exactly happens at the ultrastructural or molecular level, when two or more fungi interact on or in a plant. Only first steps were gone in this direction. An early example is the study of Weston and Dillon (1927) dealing with the interdependent reactions of the partners in wheat infected by yellow rust (*Puccinia glumarum* = *P. striiformis*) and bunt (*Tilletia tritici* = *T. caries*) simultaneously. Much more recently Willingale (1983) investigated the mutual influence of ergot (*Claviceps purpurea*) and bunt in wheat. But most workers only give phenomenological descriptions of observed effects. Such described effects can be divided into at least three categories:

(a) Antagonistic effects: The presence of a second fungus decreases the pathogenic abilities of a parasite. The relevant studies are mostly (semi-)experimental and do often not sufficiently take into consideration the natural conditions on or in the plant, especially the coenotical relations between the fungi. The purpose of such studies is to find applicable methods for biological control of economically important fungal pathogens.

(b) Synergistic effects: Combinations of two or more parasites are more harmful to the host than is either fungus alone. A recent example is the study by Wong *et al.* (1986), where the ascent of root damage in *Trifolium subterraneum* L. caused by the parasite-parasite — complex *Fusarium oxysporum* — *Phytophthora clandestina* was demonstrated, compared with the damage caused by one fungus alone.

(c) Stimulatory effects: The observed effect is a stimulation of one fungus, for instance in its ability to produce a sexual spore form. Such a case was reported by Arillaga (1935) for *Phytophthora citrophthora* when the host plant (*Citrus*) is simultaneously infected by *Diaporthe citri.* Another example is *Albugo candida,* which produces much more oospores in the tissue of *Capsella* when this is infected by *Peronospora parasitica,* too (Webster 1983).

The exploration and recognition of coenotical relations between, and defined mycosynusiae of, parasitic micromycetes on higher plants will perhaps provide a more stable base for such studies.

8.3.6. *Endophytes*

We draw attention to a special mycocoenological problem — the endophytes. Unfortunately we failed to find any definition of this term in the dictionaries and textbooks available to us, although it is occasionally used in the mycological literature. A short account on the history of the term was given by Riesen (in Riesen and Sieber 1985). We would like to introduce the following circumscription: a fungal endophyte is an individual living within another organism (plant or fungus) without any discernible negative effect. Endophytism should therefore be distinguished from parasitism in its strict sense; it can be considered as a special case of commensalism.

The studies of Müller and his pupils at ETH Zürich about this phenomenon (e.g. Petrini 1978, Petrini *et al.* 1979, Luginbühl and Müller 1980, Widler and Müller 1984, Riesen and Sieber 1985) as well as others may be summarized as follows: a great number of green terrestrial plants (perhaps nearly all) have fungal inhabitants in almost all of their parts, although they seem to be healthy and uninfected. Besides ubiquitous species there are host-specific fungi restricted to one species of green plants. The species number of endophytic communities often is high. These communities may well represent true mycosynusiae, which show characteristic differences in different parts of the plant. Positive effects for the hosts (e.g. higher resistance against parasite infections) due to the presence of endophytes can variously be assumed or were already pointed out exactly (Luginbühl and Müller 1982). The endophytic communities do not exist independently of the surface fungi, because many phylloplane species are to be found as endophytes, too. Usually the endophytic communities show a seasonal pattern of variation, which plays an important role at the beginning of decomposition of the plant tissue in question during senescence. There can be stated a true succession of the fungal endophytic community in most examined cases, which finally leads to a change into a parasitic behaviour of the fungi involved with proceeding age of the host. This is the reason for treating the endophytic fungal communities in this chapter. It is impossible to make always a clear distinction between parasitic and saprophytic micromycete communities associated with living plants.

A remarkable recent example is that given by Roberts *et al.* (1986) on the endophytic mycoflora of superficially sterilized sunflower achenes. 98 different species of fungi were growing out of the achenes, mostly microfungi but even a basidiomycete (*Schizophyllum commune* Fr.: Fr.). It seems certain that already in the near future a lot of still undiscovered fungal societies in parts of living plants will be described.

The case of the hyphomycete *Calcarisporium arbuscula* Preuss, which lives as an endophyte in apparently healthy basidiocarps of higher fungi, especially *Russula* spp. and *Lactarius* spp. (Watson 1955, 1965), somewhat represents another level. This micromycete does no harm to its host. On the contrary, the *Russula/Lactarius* fruit-body seems to be better protected against parasite attacks (e.g. by *Peckiella* spp.) if *Calcarisporium* is present. This is in agreement with experimental studies by Hirsch (1979), who was able to demonstrate high antagonistic and mycoparasitic abilities of *Calcarisporium* in vitro. However, the system *Calcarisporium* — *Russula* (*Lactarius*) seems to be balanced as long as spore production of the basidiomycete happens. Only when the fruit-body has died, *Calcarisporium conidiophores* cover the remains. What is the coenological importance of such a close coexistence of two fungi, whose relation is not a parasitic one? Could this be called a mycosynusia? Is this comparable with other fungal communities?

Fungal endophyte communities offer special problems from a mycocoenological point of view. If the kind of interaction between them and the tissues where they live will be better understood, clearer conclusions can be drawn how these consociations are to be valued.

8.3.7. *Hyperparasites and other mycoparasites*

Fungicolous fungi are commonly divided into two groups following Barnett (1963): (a) necrotrophic parasites, which kill their host fungus, and (b) biotrophic mycoparasites, showing balanced relationships. Consequently, we have to discuss two different coenological aspects: 1. the consociation of more than one necrotrophic parasite on a fungal host; 2. the valuation of stable host-parasite relations in the case of biotrophic mycoparasitism.

1. Most information available on hyperparasite consociations (which are considered as necrotrophic) are from three studies dealing mainly with tropical epiphytic fungi (Hansford 1946, Deighton 1969, Deighton and Pirozynski 1972). In these studies a lot of references to hyperparasite communities on phytoparasitic fungi can be found. However, no definite conclusions can be drawn, and a coenological analysis of the presented data is not possible. But it seems to be certain that in the case of tropical phylloplane fungi definable hyperparasite associations exist. The situation in extratropical areas is not so clear. Except the information of Blumer (1967) on associated fungi in powdery mildew colonies (saprobes and hyperparasites), nearly no information is available, and we do not have any personal experiences.

2. Biotrophic mycoparasites, especially the so-called contact parasites (e.g. Barnett 1958, Gain and Barnett 1970, etc.) were subject of different studies. They are nearly unable for axenic growth in culture, and they were found in nature practically only in connection with their hosts. This connection is so close that at least in one case (*Hansfordia parasitica* — *Physalospora obtusa*) cytoplasmatic contact at the host-parasite interface and even nucleus migrations could be observed (Hoch 1976). It is clear that these communities are not directly comparable with other fungal associations, because the partners are not at the same trophic level. On the other hand, the system host fungus — biotrophic parasite is energetically and materially balanced and more stable than any other fungal consociation.

In this chapter the authors have tried to provide a comprehensive account on mycocoenological aspects of fungal parasite communities. Not every aspect could be taken into consideration and could be treated in detail, respectively. Till now, however, such an account was not in existence. Therefore, it would be surprising if there are no mistakes, misinterpretations or misunderstandings in the text. Nevertheless, we hope to give some fruitful ideas concerning the examination of communities of parasitic microfungi.

References

Apinis, A. P. 1973. Facts and problems. Mycopathol. mycol. appl. 48: 93—109.

Arillaga, J. G. 1935. The nature of inhibition between certain fungi parasitic on *Citrus*. Phytopathology 25: 763—775.

Barkman, J. J. 1976. Allgemene inleiding tot de oecologie en sosiologie van macrofungi. Coolia 19: 57—66.

Barkman, J. J., J. Moravec, and S. Rauschert. 1986. Code of phytosociological nomenclature. 2nd edition. Vegetatio 67: 145—195.

Barnett, H. L. 1958. A new *Calcarisporium* parasitic on other fungi. Mycologia 50: 497—500.

Barnett, H. L. 1963. The nature of mycoparasitism by fungi. Ann. Rev. Microbiol. 17: 1—14.

Berkeley, M. J. 1836. Fungi II, 2. *In*: W. J. Hooker English flora, Vol. V(2).

Blakeman, J. P. (ed.). 1981. Microbial Ecology of the Phylloplane. Academic Press, London.

Blumer, S. 1933. Die Erysiphaceen Mitteleuropas mit besonderer Berücksichtigung der Schweiz. Beitr. Krypt. flora Schweiz 7: II—X, 1—483.

Blumer, S. 1967. Echte Mehltaupilze (Erysiphaceae). VEB Gustav Fischer, Jena.

Braun, U. 1978a. Phytoparasitische Pilze in den Schadzonen der Dübener Heide — Untersuchungen zur Bioindikation. Diplomarbeit, MLU Halle.

Braun, U. 1978b. Beitrag zur Floristik, Nomenklatur und Biologie phytoparasitischer Pilze. Gleditschia 6: 171—176.

Braun, U. 1982. Phytozönologisch-mykofloristische Studien über phytoparasitische Pilze in Agrarlandschaften der südlichen DDR. Diss., MLU Halle.

Buhr, H. 1956. Zur Kenntnis der Peronosporaceen Mecklenburgs. Arch. Freunde Naturgesch. Meckl. 2: 109—243 ('1955/56').

Burdon, J. J. 1982. The effect of fungal pathogens on plant communities. p. 99—112. *In*: E. I. Newman (ed.) The Plant Community as a Workable Mechanism. Oxford.

Burdon, J. J., and R. C. Shattock. 1980. Disease in plant communities. Appl. Biol. 5: 145—219.

Chilvers, G. A., and E. G. Brittain. 1972. Plant competition mediated by host-specific parasites — A simple model. Austr. J. Biol. Sci. 25: 749—756.

Cummins, G. B. 1979. Annotated, illustrated, host index of Sonoran Desert rust fungi. Mycotaxon 10: 1—20.

Deighton, F. C. 1969. Microfungi IV. Some hyperparasitic hyphomycetes and a note on *Cercosporella uredinophila* Sacc. Mycol. Papers 118: 1—41.

Deighton, F. C., and K. A. Pirozynski. 1972. Microfungi V. More hyperparasitic hyphomycetes. Mycol. Papers 128: 1—110.

Dennis, R. W. G., D. A. Reid, and B. M. Spooner. 1977. The Fungi of the Azores. Kew Bull. 32: 85—136.

Dinoor, A., and N. Eshed. 1984. The role and importance of pathogens in natural communities. Ann. Rev. Phytopathol. 22: 443—466.

Dörfelt, H. 1974a. Mykofloristische, mykocoenologische und mykogeographische Studien in Naturschutzgebieten mit Xerothermstandorten im Süden der DDR unter besonderer Berücksichtigung der Gebiete Leutratal, Steinklöbe und Neue Göhle. Diss., MLU Halle.

Dörfelt, H. 1974b. Die Erforschung der Mykozönosen als Elemente der Ökosysteme. Mitt. Sekt. Geobot. Phytotax. Biol. Ges. DDR, Sonderheft 'Grundlagen der Ökosystemforschung': 85—91.

Dörfelt, H., and U. Braun. 1977. Beachtenswerte Funde phytoparasitischer Pilze in der DDR (I). Hercynia N.F. 14: 11—20.

Durrieu, G. 1966, Étude écologique de quelques groupes de champignons parasites des plantes spontanées dans les Pyrénées. Bull. Soc. Hist. Nat. Toulouse 102: 7—277.

Ellenberg, H. 1978. Vegetation Mitteleuropas mit den Alpen (2nd ed.) Ulmer, Stuttgart.

Gain, R. E., and H. L. Barnett. 1970. Parasitism and axenic growth of the mycoparasite *Gonatorhodiella highlei*. Mycologia 62: 1122—1129.

Gäumann, E. 1923. Beiträge zu einer Monographie der Gattung *Peronospora* Corda. Beitr. Krypt. flora Schweiz 5: 1—360.

Geljuta, V. P. 1979a. Porivnjal'ne vivčennja flori Erysiphaceae stepovich zapovidnikiv URSR. Ukr. Bot. Ž. 36: 586—590.

Geljuta, V. P. 1979b. Borošnisto-rosjani gribi (Erysiphaceae) u fitocenozach zapovidnika 'Proval's'kij Step'. Ukr. Bot. Ž. 36: 476—478.

Greville, R. K. 1824. Flora Edinensis. Blackwood, Edinburgh.

Gustavsson, A. 1959. Studies on Nordic Peronosporas. I. Taxonomic Revision. Opera Bot. 3: 1—271.

Hansford, C. G. 1946. The foliicolous ascomycetes, their parasites and associated fungi. Mycol. Papers 15: 1—240.

Harlan, J. R. 1976. Diseases as a factor in plant evolution. Ann. Rev. Phytopathol. 14: 31—51.

Hirsch, G. 1979. Vergleichende Untersuchungen zur Biologie von *Calcarisporium, Acrodontium* und *Tritirachium* in Reinkultur. Diplomarbeit, FSU Jena.

Hirsch, G. 1982. Die autözischen *Uromyces*-Arten (Basidiomycetes, Uredinales) auf *Euphorbia seguierana* Necker. Wiss. Z. Univ. Jena. Math.-Nat. R. 31: 229—238.

Hirsch, G. 1985. Die Brandpilze Thüringens — Nachtrag I. Haussknechtia 1: 43—49 ('1984').

Hoch, H. C. 1976. Mycoparasitic relationships. III Parasitism of *Physalospora obtusa* by *Calcarisporium parasiticum*. Can. J. Bot. 55: 198—207.

Kreisel, H. 1957. Die Pilzflora des Darß und ihre Stellung in der Gesamtvegetation. Feddes Repert. Beih. 137: 110—183.

Kreisel, H. 1985. Pilzsoziologie. p. 67—82. *In*: Michael—Hennig—Kreisel, Handbuch für Pilzfreunde Vol. IV (3rd edition). VEB Gustav Fischer, Jena.

Kreisel, H., and K.-H. Müller. 1987. Das Pleurotetum cornucopiae, eine Pilzgesellschaft an toten Ulmenstämmen im Gefolge des Ulmensterbens. Arch. Naturschutz Landschaftsforsch. 27: 17—25.

Kućmierz, J. 1973. Grzyby pasożytnicze w ziborowiskach roślinnych Ojcowskiego Parku Narodowego. Ochrona Przyrody 38: 155—211.

Kućmierz, J. 1977. Studia nad grzybami fitopatogenicznymi z Pienin. Zeszyty Nauk. Akad. Roln. Krakowie, Rozpr. 52: 3—142.

Leppik, E. E. 1970. Gene centers of plants as sources of disease resistance. Ann. Rev. Phytopathol. 8: 323—344.

Luginbühl, M., and E. Müller. 1980. Endophytische Pilze in den oberirdischen Organen von 4 gemeinsam an gleichen Standorten wachsenden Pflanzen (*Buxus, Hedera, Ilex, Ruscus*). Sydowia 33: 185—209.

Luginbühl, M., and E. Müller. 1982. Untersuchungen über endophytische Pilze II. Förderung der Samenkeimung bei *Hedera Helix* durch *Aureobasidium pullulans* und *Epicoccum purpurascens*. Ber. Schweiz. Bot. Ges. 90: 262—267.

Luttrell, F. A. 1974. Parasitism of fungi on vascular plants. Mycologia 66: 1—15.

Magnus, P. 1894. Das Auftreten der *Peronospora parasitica*, beeinflußt von der Beschaffenheit und dem Entwickelungszustande der Wirthspflanze. Ber. Dt. Bot. Ges. 12: 39—44.

Majewski, T. 1967. Przyczynek do flory grzybów pasożtniczych Puszczy Kampinoskiej. Acta Mycol. 3: 115—151.

Majewski, T. 1971. Grzyby pasożytnicze Bialowieskiego Parku Narodowego na tle mikoflory Polski. (Peronosporales, Erysiphaceae, Uredinales, Ustilaginales). Acta Mycol. 7: 299—388.

Mäkinen, Y. 1966. On the macroecology of some rust fungi. Ann. Univ. Turku A2, 36: 75—84.

Mułenko, W. 1981. Badania nad mikroskopijnymi grzybami pasożytniczymi rezervatu leśnoturfowiskowego Brzeziczno. Ann. Univ. M.C.S. Lublin, Polonia, sect. C, 36: 81—88.

Parmelee, J. A. 1984. Microfungi parasitic on vascular plants in Waterton Lake National Park, Alberta, and environs. Agric. Canada, Res. Branch, Techn. Bull. 1984-11E: 1—32.

Parmelee, J. A., and J. Ginns. 1986. Parasitic microfungi on vascular plants in the Yukon and environs. Int. J. Mycol. Lichenol. 2: 293—347.

Petrak, F. 1927. Mykologische Notizen. Ann. Mycol. 25: 193—343.

Petrini, O. 1978. Untersuchungen über endophytische Pilze von *Juniperus communis* L. Diss., ETH Zürich.

Petrini, O., E. Müller, and M. Luginbühl. 1979. Pilze als Endophyten von grünen Pflanzen. Naturwissenschaften 66: 262.

Poelt, J. 1981. Biotroph-parasitische Frühlingspilze aus dem mittleren und südlichen Burgenland. Natur Umwelt Burgenland 4: 57—63.

Riesen, T., and T. Sieber. 1985. Endophytic fungi in winter wheat (*Triticum aestivum* L.). Endophytische Pilze von Winterweizen (*Triticum aestivum* L.). Diss., ETH Zürich.

Roberts, R. G., J. A. Robertson, and R. T. Hanlin. 1986. Fungi occuring in the achenes of sunflower (*Helianthus annuus*). Can. J. Bot. 64: 1964—1971.

Romaszewska-Sałata, J. 1976. Grzyby pasożytnicze zbiorowisk Stepowych na Wyżynie Lubelskiej (Peronosporales, Erysiphales, Uredinales, Ustilaginales). Folia Soc. Sci. Lubl. 18, Biol. 2: 91—96.

Romaszewska-Sałata, J. 1977. Grzyby pasożytnicze zbiorowisk stepowych na Wyżynie Lubelskiej. Acta Mycol. 13: 25—83.

Romaszewska-Sałata, J. 1981. Materiały do poznania mikroskopijnych grzybów fitopatogenicznych zbiorowisk kserotermicznych na Wyżynie Małopolskiej. Ann. Univ. M.C.S. Lublin — Polonia, sect. C, 36: 51—69.

Rothmaler, W. 1982. Exkursionsflora für die Gebiete der DDR und der BRD. Band 4. Kritischer Band. (5th edition edited by R. Schubert and W. Vent). Volk und Wissen, Berlin.

Sandu-Ville, C., E. Eliade, I. Comes *et al.* 1973. Micoflora din zona sistemului Hidroenergetic si de navigatie 'Portile de Fier' — Romania. Acta Bot. Horti Bucurest. 1972—73: 579—645.

Savile, D. B. O. 1971. Co-ordinated studies of parasitic fungi and flowering plants. Naturalistes Canad. 98: 535—552.

Savile, D. B. O., and Z. Urban. 1982. Evolution and ecology of *Puccinia graminis*. Preslia 54: 97—104.

Scholz, H. 1976. Veränderungen der Berliner Brandpilzflora. Schriftenr. Vegetationsk. 10: 215—225.

Sörgel, G. 1957. Vorkommen und Verbreitung epiphyller Pilze in China. Z. Pilzk. 23: 100—117.

Sutton, B. C. 1986. Presidential Address. Improvizations on conidial themes. Trans. Brit. Mycol. Soc. 86: 1—38.

Tanda, S. 1983. Mycoflora in the Okutama Experimental Forest of Tokyo University of Agriculture and its environs. J. Agr. Sci. Tokyo Nogyo Daigaku 27: 214—232.

Thomas, M. R., and R. C. Shattock. 1986. Filamentous fungal associations in the phylloplane of *Lolium perenne*. Trans. Brit. Mycol. Soc. 87: 255—268.

Tutin, T. G. *et al.* 1964—1980. Flora Europaea. Vol. 1—5. Cambridge University Press, Cambridge.

Urban, Z. 1980. Rust ecology and phytocoenology as aids in rust taxonomy. Rep. Tottori Mycol. Inst. 18: 269—273.

Vánky, K. 1985. Carpathian Ustilaginales. Symb. Bot. Upsal. 24(2).

Watson, P. 1955. *Calcarisporium arbuscula* living as an endophyte in apparently healthy sporophores of *Russula* and *Lactarius*. Trans. Brit. Mycol. Soc. 38: 409—414.

Watson, P. 1965. Further observations on *Calcarisporium arbuscula*. Trans. Brit. Mycol. Soc. 48: 9—17.

Webber, J. F., and J. N. Hedger. 1986. Comparison of interactions between *Ceratocystis ulmi* and elm bark saprobes in vitro and in vivo. Trans. Brit. Mycol. Soc. 86: 93—101.

Webster, J. 1983. Pilze — Eine Einführung. Springer, Berlin, Heidelberg, New York.

Weston, W. A., and R. Dillon. 1927. The incidence and intensity of *Puccinia glumarum* on wheat infected and noninfected with *Tilletia tritici*, showing an apparent relationship between rust and bunt. Ann. Appl. Biol. 14: 105—112.

Widler, B., and E. Müller. 1984. Untersuchungen über endophytische Pilze von *Arctostaphylos uva-ursi* (L.)Sprengel (Ericaceae). Botanica Helvetica 94: 307—337.

Willingale, J. 1983. Biochemical interactions involving parasitic *Claviceps purpurea*. Ph.D. Thesis, University of London.

Wong, D. H., K. Sivasithamparam, and M. J. Barbetti. 1986. Influence of soil temperature, moisture and other fungal root pathogens on pathogenicity of *Phytophthora clandestina* to subterranean clover. Trans. Brit. Mycol. Soc. 86: 479—482.

Index

GPSR Compliance

The European Union's (EU) General Product Safety Regulation (GPSR) is a set of rules that requires consumer products to be safe and our obligations to ensure this.

If you have any concerns about our products, you can contact us on ProductSafety@springernature.com

In case Publisher is established outside the EU, the EU authorized representative is:

Springer Nature Customer Service Center GmbH
Europaplatz 3
69115 Heidelberg, Germany

Batch number: 10816188

Printed by Printforce, the Netherlands